# NATURAL DEDUCTION

## A Proof-Theoretical Study

### DAG PRAWITZ

*Department of Philosophy*
*Stockholm University*

**DOVER PUBLICATIONS, INC.**
Mineola, New York

*Bibliographical Note*

This Dover edition, first published in 2006, is an unabridged republication of the work originally published in the "Stockholm Studies in Philosophy" series by Almqvist & Wiksell, Stockholm, in 1965. A new Preface to the Dover Edition, written by the author, has been specially prepared for the present volume. An errata list has also been added.

*Library of Congress Cataloging-in-Publication Data*

Prawitz, Dag.
    Natural deduction : a proof-theoretical study / Dag Prawitz.
        p. cm.
    Originally published: Stockholm : Almqvist & Wiksell, 1965.
    Includes bibliographical references and index.
    ISBN 0-486-44655-7 (pbk.)
        1. Logic. 2. Logic, Symbolic and mathematical. 3. Modality (Logic)
    4. Gentzen, Gerhard. I. Title.

BC71.P68 2006
162—dc22
                                                           2005051968

Manufactured in the United States of America
Dover Publications, Inc., 31 East 2nd Street, Mineola, N.Y. 11501

# PREFACE TO THE DOVER EDITION

This monograph was first published in 1965 as my doctoral dissertation. For a doctoral thesis, it was printed in a fairly large number of copies, but it was nevertheless quickly sold out. The question then arose whether to reprint the book. My reason for not doing so was the many new results that were soon obtained on the basis of the scheme presented in my monograph. It seemed to me that they deserved a new and more comprehensive presentation, which should contain the results of my monograph as a first part. Rather than republish the monograph, I should therefore write such a new book.

As a first step toward this, I summarized the main topics of the monograph and surveyed some of the new results in a paper "Ideas and Results in Proof Theory," which appeared in *Proceedings of the Second Scandinavian Logic Symposium* in 1971 (North-Holland Publishing Company). But the field continued to grow (for instance, in the very same Proceedings new results were presented by Per Martin-Löf and Jean-Yves Girard). Soon afterwards I got involved in other projects as well, among them applications of the ideas of the dissertation to the philosophy of language. And so far the new book has not been written.

Therefore, when Dover Publications recently proposed reprinting my monograph, I finally agreed. This does not mean that I have completely given up the idea of writing a more comprehensive book, but only that I now realize that such a book would be quite different from the original monograph, which is now reprinted without change except for a list of errata (almost the same list was inserted as a loose leaf in the first edition three years after its publication).

The main theme of the monograph is the establishing of a certain normal form for derivations in the systems of natural deduction that were introduced by Gentzen in his doctoral dissertation "Untersuchungen über das logische Schliessen" in 1934. The result, that every derivation in a system of natural deduction can be transformed to this normal form by certain reductions defined in the monograph, is known as the *Normalization Theorem* for the system in question, a term commonly used since the time of the above-mentioned Proceedings (following a suggestion by Georg Kreisel). It is equivalent to what is known as the *Hauptsatz* for Gentzen's corresponding calculi of sequents. The two results can easily be obtained from one another as is shown in the monograph. However, as claimed in the Preface to the first edition of my monograph, the result for natural deduction is "in many ways simpler and more illuminating." In the main text I go on to describe what I call an inversion principle, which is said to underlie the central idea of Gentzen's Hauptsatz, but which more directly gives rise to the Normalization Theorem that I prove for systems of natural deduction.

Gentzen would certainly have agreed with my remark that the result for natural deduction is more illuminating. When he presents the underlying idea of his consistency proof for arithmetic in "Neue Fassung des Widerspruchsfreiheitsbeweises für die reine Zahlentheorie" he gives what is virtually an intuitive, though incomplete sketch of the Normalization Theorem, and in a footnote he remarks that "moreover, exactly the same reasoning underlies the proof of the Hauptsatz of my dissertation." In the synopsis of his dissertation he had already remarked that the system of natural deduction contains "the properties essential to the validity of the Hauptsatz," but "only in its intuitionist form." When presenting the consistency proof, he says that the special position of negation, which in the natural system for classical logic constitutes "a troublesome exception," has been completely removed in the sequent calculus "as if by magic." He adds: "I myself was greatly surprised at this property of the sequent calculus, when I first constructed it."

It is thus obvious that Gentzen obtained his results by pondering upon natural deductions and that insights concerning them inspired his formal work, which he carried out in published writings only for the calculi of sequents owing to problems he met in the case of classical logic when trying to formulate his insights directly for natural deductions. But it is not obvious from his published writings to what extent he actually obtained results for natural deductions. It is therefore of great historical interest that an early handwritten draft of Gentzen's dissertation has been discovered, which throws light upon this question. The draft had lain dormant among papers left by Paul Bernays. The existence of this early draft has been known for a long time, because one of its pages is actually reproduced photographically in the English translation of Gentzen's papers, *The Collected Papers of Gerhard Gentzen,* edited by M. E. Szabo (North-Holland Publishing Company 1969). But the content of the draft was not generally known. Only a few months ago Professor Jan von Plato found in it a detailed proof of the Normalization Theorem for the intuitionist system of natural deduction. This finding will be described by von Plato in a paper "Gentzen's Logic" to appear in volume V of the *Handbook of the History and Philosophy of Logic*, which is to be published by North-Holland Elsevier in 2005. It turns out that Gentzen's proof is based on substantially the same kind of arguments as my proof in this monograph. There is no indication of a similar result for classical logic, although as is seen in the present monograph, when a system of natural deduction for classical logic is formulated appropriately, a proof of a Normalization Theorem is even simpler than in the intuitionist case. Further surprises may come, however, because additional manuscripts of Gentzen have come to light. They are in shorthand and are being transcribed by Professor Christian Thiel for possible publication, perhaps together with the above-mentioned manuscript from the Bernays archive.

Gentzen's chief scientific aim as he formulated it himself as early as 1932 was to contribute to Hilbert's program.[1] Part of his great achievement in that respect, which changed proof theory and has had a lasting influence on the field, depended on general insights concerning the structure of proofs. To my mind, these insights are at least as interesting as his contributions to Hilbert's program. They consisted first of all in a specific logical analysis of informal proofs that resulted in the systems of natural deduction. The significance of this result is stated here and there in Gentzen's published writings, mostly in informal comments as exemplified above, but it becomes obscured in his formal work concerned with the calculi of sequents. The present monograph is a first attempt to bring out this significance more clearly.

Saying this, I want in no way to downplay the importance of the calculi of sequents. They have a significance in their own right, particularly in connection with classical model theory, as was brought out in a new style of completeness proof for first order predicate logic introduced in the 1950s by Beth, Hintikka, Kanger, and Schütte.[2]

---

[1] See letter quoted by Menzler-Trott in his book, *Gentzens Problem*, Birkhäuser, Basel 2001, page 35. This aim is of course also clear in Gentzen's published writings from 1934 and on.

[2] I try to give a systematic account of this aspect of the sequent calculus for classical logic in the paper "Comments on Gentzen-Type Procedures and the Classical Notion of Truth" in *Proof Theory Symposium Kiel 1974*. Lecture Notes in Mathematics 500, Berlin 1975.

Gentzen did not concern himself with these aspects in his published writings, and they do not occur in the present monograph either. One may note here that for some time it was only by use of this connection that we were able to extend the Normalization Theorem to impredicative higher order logics in a weaker form, namely, that to every proof there exists a normal one for the same conclusion from the same hypotheses (though not got by reductions as in first order logic).

The analysis of inferences by Gentzen that resulted in his system of natural deduction is especially relevant in a constructive approach to logic. In view of the fact that classical logic sometimes can be imbedded in intuitionist logic, or that the addition of a classical rule to intuitionist logic is fairly harmless in some contexts, it yields results for classical logic too. But in philosophy of logic the theme of this monograph has been fruitful in particular for a constructive direction of the field, as witnessed in later works by, among others, Michael Dummett, Per Martin-Löf, and myself. These philosophical applications are only hinted at in this book.

Stockholm, May 2005                              *Dag Prawitz*

# PREFACE

Systems of natural deduction, invented by Jáskowski and by Gentzen in the early 1930's, constitute a form for the development of logic that is natural in many respects. In the first place, there is a similarity between natural deduction and intuitive, informal reasoning. The inference rules of the systems of natural deduction correspond closely to procedures common in intuitive reasoning, and when informal proofs—such as are encountered in mathematics for example—are formalized within these systems, the main structure of the informal proofs can often be preserved. This in itself gives the systems of natural deduction an interest as an explication of the informal concept of logical deduction.

Gentzen's variant of natural deduction is natural also in a deeper sense. His inference rules show a noteworthy systematization, which, among other things, is closely related to the interpretation of the logical signs. Futhermore, as will be shown in this study, his rules allow the deduction to proceed in a certain direct fashion, affording an interesting normal form for deductions. The result that every natural deduction can be transformed into this normal form is equivalent to what is known as the *Hauptsatz* or the *normal form theorem*, a basic result in proof theory, which was established by Gentzen for the calculi of sequents. The proof of this result for systems of natural deduction is in many ways simpler and more illuminating.

Gentzen's systems of natural deduction are developed in detail in Chapter I. Their characteristic features can be described roughly as follows. With one exception, the inference rules are of two kinds, viz., *introduction rules* and *elimination rules*. The properties of each sentential connective and quantifier are expressed by two rules, one of each kind. The introduction rule for a logical constant allows an inference *to* a formula that has the constant as principal sign. For instance, the introduction rule for & (and) allows the inference to $A \& B$ given the two premisses $A$ and $B$, and the introduction rule for

⊃ (if—, then—) allows the inference to $A \supset B$ given a deduction of $B$ from $A$. The elimination rule for a constant, on the other hand, allows an inference *from* a formula that has the constant as principal sign. For instance, the elimination rule for & allows an inference from $A \& B$ to $A$ (or to $B$), and the elimination rule for ⊃ allows the inference from the premisses $A$ and $A \supset B$ to $B$.

For inferring certain formulas, the introduction rule gives thus a sufficient condition that is formulated in terms of subformulas of these formulas. The elimination rule, on the other hand, is related to the corresponding introduction rule according to a certain *inversion principle*: the elimination rule is in a sense only the inverse of the corresponding introduction rule.

This inversion principle will be described in more detail in Chapter II, and by its use the result mentioned above that every deduction can be brought to a certain normal form is easily obtained. A deduction in normal form proceeds from the assumptions of the deduction to the conclusion in a direct and rather perspicuous way without detours; roughly speaking, the assumptions are first broken down into their parts by successive applications of the elimination rules, and these parts are then combined to form the conclusion by successive applications of the introduction rules. This is established for classical logic in Chapter III and for intuitionistic and minimal logic in Chapter IV. Applications of the result are exemplified by some corollaries.

The systems are extended in Chapter V to incorporate second order logic also. The main results for first order logic are then extended to ramified second order logic.

In Chapter VI the results are extended in another direction to the modal systems known as S4 and S5, and in Chapter VII, to some systems based upon concepts of implication that differ from the usual ones. Among the concepts considered is one introduced by Church and another one introduced by Ackerman (studied especially by Anderson and Belnap).

Some related systems are taken up in three appendices. The relation between the systems of natural deduction and the calculi of sequents is described in Appendix A, and Gentzen's Hauptsatz is then obtained as a corollary. A demonstrably consistent set-theory developed by Fitch is discussed in Appendix B. The final appendix is devoted to some other systems of natural deduction that have been described in the literature. Among them are the first systems developed by Jaskowski, some variants of Gentzen's systems of natural deduction, and systems containing rules for existential instantiation.

\*

The main results concerning a normal form for natural deduction was presented at joint colloquiums at the Universities of Uppsala and Stockholm in 1961 and—in a more complete form—at two colloquiums at the University of Münster (Westfalen) in 1962. An abstract of it was presented at a meeting of the Association for Symbolic Logic in New York 1964. The material in Chapters II and IV is presented essentially as at these colloquiums. Some of the corollaries, however, are of a somewhat later date, and so is the present treatment of classical logic in Chapter III. Most of the content in Chapters VI and VII was included in seminars that I held at UCLA during the spring semester of 1964.

I am indebted to participants in these colloquiums and seminars for criticism and discussion. I am especially grateful to Professors Anders Wedberg, Stig Kanger, and Christer Lech, who read parts of the manuscript and made valuable suggestions. I also want to thank Fil.lic. Göran Enger for help with proof-reading, and Mrs. Muriel Bengtsson, Mr. John Swaffield, and Mr. Rolf Schock for checking parts of the manuscript with respect to the English.

*Dag Prawitz*

# ERRATA

Three more important corrections are listed first:

1. In Corollary 7 on p. 56, it shall be assumed that no operational constant occurs in $\exists xA$ or in any formula of $\Gamma$. Without this assumption the corollary is incorrect. Note also the misprint in clause (ii) of the corollary where "$\Gamma$" is to be inserted before the second occurrence of "$\vdash$".

The proof on p. 57 is defective when beginning on line 7, it is said "If $u$ is not one of $t_1, t_2, \ldots, t_n$, we now replace all occurrences of $u$ in $\Pi$ by one of the terms $t_1 \ldots$". If $u$ contains no operational constant, the replacement is in order, but otherwise such operational constants must first be removed throughout the deduction.

In general, an operational constant that occurs neither in the endformula nor in any undischarged assumption can be left out throughout the deduction without affecting its validity. With this amendment, the proof also establishes the following stronger result

COROLLARY $7'$. (i) Provided no formula in $\Gamma$ has a strictly positive subformula with principal sign $\exists$: If $\Gamma \vdash \exists xA$, then $\Gamma \vdash A_{t_1}^x \lor A_{t_2}^x \lor \ldots \lor A_{t_n}^x$ for some terms $t_1, t_2, \ldots, t_n$ built up of symbols occurring in $\exists xA$ or in some formula of $\Gamma$.

(ii) Provided no formula in $\Gamma$ has a strictly positive subformula with principal sign $\exists$ or $\lor$: If $\Gamma \vdash \exists xA$, then $\Gamma \vdash A_t^x$ for some term $t$ built up of symbols occurring in $\exists xA$ or in some formula of $\Gamma$.

2. In the definition of *rank* on page 69, line 10 f. b. replace "containing a bound variable of level $m-1$" by "containing a bound variable of level $m-1$ or a parameter or constant of level $m$".

3. In the first paragraph on page 59, the last three sentences (beginning "However") are incorrect and shall be deleted. In fact, it can be seen that weak definability implies strong definability (see footnote 2 in PRAWITZ–MALMNÄS, A survey of some connections between classical, intuitionistic, and minimal logic, in: *Contributions to mathematical logic* (North-Holland Publ. Co.), pp. 215–229, Amsterdam 1968. This fact could have been used to simplify the proof of Corollary 9 slightly.

Other corrections:

Page

| | | |
|---|---|---|
| 1 | Line 1 | for "Jáskowski" read "Jaśkowski" (this error is recurrent) |
| 12 | Line 4 | for "explicitely" read "explicitly" |
| 13 | Line 2 | for "Introduction" read "Preface" |
| 15 | Footnote | for "required" read "require" |
| 19 | Line 14 | for "$\forall x \forall y (Pvy \supset Pyx)$" read "$\forall x \forall y (Pxy \supset Pyx)$" |
| 23 | Line 21 | for "$\lor$ E" read "$\lor$E" |
| 26 | Line 14 f. b. | for "treee" read "tree" |
| 28 | Line 8 f. b. | for "$\Pi_1$" read "$\Pi_1 = (\sum/A)$" |
| 28 | Line 5 f. b. | for "$(\Pi_1/[A]/\Pi_2)$" read "$(\sum/[A]/\Pi_2)$" |

Page

| 28 | Line 2 f. b. | insert "$a$" after "parameter" |
|---|---|---|
| 29 | Line 6 | for "chose" read "choose" |
| 29 | Line 9 f. b. | for "appliation" read "application" |
| 42 | Line 7 | for "segment" read "formula" |
| 45 | Footnote, line 5 f. b. | delete sentence beginning "The result" |
| 45 | Footnote, line 2 f. b. | for "Corollary 1" read "Corollary 4" |
| 49 | Line 10 f. b. | for second occurrence of "$\sigma_1$" read "$\sigma_2$" |
| 51 | Uppermost figures | replace all occurrences of "$C$" except those in the formula "$B \vee C$" with "$D$" |
| 51 | Line 6 f. b. | for "right" read "left" |
| 53 | Line 16 | for "$\sigma_j$" read "$\sigma_{j+1}$" |
| 53 | Line 17 | for "$\sigma_{j+1}$" read "$\sigma_j$" |
| 54 | Line 10 | for "CORALLARY" read "COROLLARY" |
| 55 | Line 9 | for "Corallary" read "Corollary" |
| 56 | Line 2 f. b. | for "terms $t_1$, $t_2$, ..., $t_n$" read "parameters" |
| 57 | Line 4 | for "positve" read "positive" |
| 59 | Line 10 | for "CORALLARY" read "COROLLARY" |
| 67 | Footnote 2, line 2 | after "definition" insert "of $A \vee B$ as $\forall X((A \supset X)$ & $(B \supset X) \supset X)$ and" |
| 69 | Line 11 f. b. | for "consatnts" read "constants" |
| 76 | Line 7 | for "indentical" read "identical" |
| 84 | Uppermost figure to the left | the lower occurrences of "(1)" and "(2)" are to be interchanged |
| 88 | Footnote 1, line 8 | for "intuionistic" read "intuitionistic" |
| 90 | Line 1 | delete "which now hold as derived rules" and insert at the end of Remark 1 "The equivalences between the calculi obtained in this way and the calculi defined above are easily seen" |
| 91 | Footnote | the footnote is to be numbered "1", not "3" |
| 92 | Line 2 | for "E-rules" read "∃-rules" |
| 98 | Line 2 and 8 | for "Łukasiewics" read "Łukasiewicz" |
| 103 | end of last paragraph | it should have been pointed out that the system in Copi [2] is unsound; a way to make it correct is described in Prawitz, A note on existential instantiation, *JSL*, vol. 32 (1967), pp. 81–82. |
| 103 | footnote 1 | the footnote is to refer to the sentence ending at line 5 f. b. (and not to the one ending at line 4 f. b.) |
| 106 | Line 4 | for "Comleteness" read "Completenes" |
| 106 | Line 3 f. b. | for " Frech" read "French" |
| 108 | Line 6 | for "matemathematics" read "metamathematics" |
| 108 | Line 10 | for "Willhelm" read "William" |
| 109 | Line 3 | for "Patrik" read "Patrick" |
| 111 | Normal deductions | for "34" read "39" |

# ADDITIONS TO THE LIST OF ERRATA

Page

| | | |
|---|---|---|
| 67 | Line 1 f. b. | for "not occur not occur" read "not occur" |
| 72 | Footnote | the footnote is correct as it stands but is misleading since the announced proof was never carried out |
| 101 | Line 7 f. b. | for "1937" read "1936" |
| 101 | Footnote 3, last line | for "Jákowski" read "Jaśkowski" |
| 106 | Line 17 f. b. | insert "I" at the end of the title |
| 108 | Line 10 | for "Brittish" read "British" |
| 108 | Line 6 f. b. | for "BERKLEY" read "BARKLEY" |
| 110 | | for "Belnarp" read "Belnap" |

# CONTENTS

# NATURAL DEDUCTION

A Proof-Theoretical Study

SOME NOTATIONAL CONVENTIONS. The letters $i$, $j$, $k$, $m$, and $n$ are used as numerical indices. They are usually supposed to range over the positive integers. Sometimes the range is to include o also, which will then—if not obvious from the context—be explicitly stated.

I sometimes give two definitions in one sentence by using square brackets to indicate two readings of the sentence.

A name followed by a numeral within square brackets is a reference to the bibliography at the end of the monograph.

The number of a theorem refers to its order within the chapter; the same holds for the corollaries.

# I. NATURAL DEDUCTION OF GENTZEN-TYPE

A system of natural deduction can be thought of as a set of rules (of the "natural" kind suggested in the Introduction) that determines the concept of deduction for some language or set of languages. Together with a language such a system can thus be said to constitute a logical calculus.[1]

In this chapter, I shall describe systems of natural deduction for classical, intuitionistic, and minimal predicate logic (of first order). The systems, which essentially are of a kind introduced by Gentzen [3], will be called *Gentzen-type systems of natural deduction*. I begin by describing the languages that the systems are intended for (§ 1), and the inference rules and the deductions of the systems (§ 2).

## § 1. The languages of first order

The languages considered are languages of first order predicate logic formulated in the usual way. However, I follow the convenient practice (which is not always adopted) of using two kinds of signs that range over the individuals; one kind, called *variables*, is only used bound, and the other, called *parameters*, is only used free. The languages are to contain (primitive) signs for conjunction (&), disjunction (∨), implication (⊃), universal quantification (∀), existential quantification (∃), and falsehood (or absurdity) (⋏). In more detail, these languages (referred to by the letter L) are described as follows.

SYMBOLS. Every language L is to contain non-descriptive symbols divided into categories as follows:

---

[1] Cf. note 1 on p. 23.

1. *Individual variables.* There are to be denumerably many of this kind. As a syntactic notation to refer to arbitrary symbols of this kind, I use the letters: $x, y, z$.
2. *Individual parameters.* Denumerably many. Syntactic notation: $a$, $b, c$.
3. *$n$-place predicate parameters, $n = 0, 1, 2, \ldots$.* Denumerably many for each $n$. Syntactic notation: $P^n, Q^n$; or, when the number of places is arbitrary or clear from the context: $P, Q$.
4. *Logical constants.*
    (a) A sentential constant for falsehood (absurdity): $\curlywedge$.
    (b) Sentential connectives: &, v, and $\supset$.
    (c) Quantifiers: $\forall$ and $\exists$.
5. *Auxiliary signs:* Parantheses.

In addition a language L may contain a set of descriptive constants divided into the following categories:

6. *Individual constants.*
7. *$n$-place operation constants, $n = 1, 2, \ldots$*
8. *$n$-place predicate constants, $n = 0, 1, \ldots$*

All the different categories are to be disjoint. Certain strings of symbols are singled out as terms and formulas in the usual way.

TERMS. $t$ is an *individual term* in a language L if and only if $t$ is a (finite) symbol string and either

  1) $t$ is an individual parameter or an individual constant in L, or
  2) $t = f(t_1 t_2 \ldots t_n)$, where $t_1, t_2, \ldots, t_n$ are individual terms in L and $f$ is an $n$-place operation constant in L.

FORMULAS. $A$ is an *atomic formula* in L if and only if either (1) $A$ is $\curlywedge$, or (2) $A$ consists of an $n$-place predicate parameter or constant in L followed by $n$ individual terms in L.

The notion of *formula* in L is defined inductively by:[1]
  1) An atomic formula in L is a formula in L.
  2) If $A$ and $B$ are formulas in L, then so are $(A \mathbin{\&} B)$, $(A \lor B)$, and $(A \supset B)$.

---

[1] For perspicuity, I shall usually give inductive definitions by clauses as above. It will always be possible to state them more accurately in the form of equivalences (as in the definition of terms above).

3) If $A$ is a formula in L, then so are $\forall x A^*$ and $\exists x A^*$, where $A^*$ is $A$ or is obtained from $A$ by replacing occurrences of an individual parameter with the variable $x$.

I sometimes omit outer parentheses around a formula and parentheses in repeated conjunctions and disjunctions. Parantheses may also be omitted under the convention that $\supset$ (and $\equiv$ introduced below) makes a greater break than & and v.

The *degree* of a formula $A$ is defined as the number of occurrences of logical constants in $A$ except $\wedge$.

Occasionally there are reasons to consider also symbol strings that are like terms or formulas except for containing variables at places where a term or formula has parameters; they will be called *pseudo-terms* and *pseudo-formulas*. Formulas and terms are considered to be special cases of pseudo-formulas and pseudo-terms.

Let $A$ be a pseudo-formula that is not atomic. Then, $A$ has exactly one of the forms $(B \mathbin{\&} C)$, $(B \mathbin{v} C)$, $(B \supset C)$, $\forall x B$, and $\exists x B$; the symbol &, v, $\supset$, $\forall$, or $\exists$, respectively, is said to be the *principal sign* of $A$.

The *scope* of a certain occurrence of a logical constant in a pseudo-formula $A$ is the part of $A$ that has this occurrence as principal sign.

An occurrence of a variable $x$ in a pseudo-formula $A$ is *bound* or *free* according as the occurrence belongs or does not belong to the scope of a quantifier that is immediately followed by $x$. In a formula, all occurrences of variables are obviously bound.[1]

A *universal [existential] closure* of a formula $A$ with respect to the parameters $a_1, a_2, ..., a_n$ is to be a formula of the form $\forall x_1 \forall x_2 ... \forall x_n A^*$ $[\exists x_1 \exists x_2 ... \exists x_n A^*]$, where the variables $x_1, x_2, ..., x_n$ are all different from each other and from variables occurring in $A$, and $A^*$ is obtained from $A$ by replacing each occurrence of $a_i$ by $x_i$ $(i \leqslant n)$. A formula in which there are no parameters is a *sentence*.

SOME NOTATIONS. Notations like $A_x^t$ and $A_u^t$, where the letter $A$ represents a formula and the letters $t$ and $u$ terms, are used to denote the result of substituting $x$ or $u$, respectively, for all occurrences of $t$ in $A$ (if any). In contexts where a notation like $A_x^t$ is used, it is always to be assumed that $t$ does not occur in $A$ within the scope of a quantifier that is immediately followed by $x$. A notation like $A_i^x$, where $A$

---

[1] The distinction between free and bound occurrences of variables can be left out entirely if we required in clause 3 in the definition of formula that $x$ not occur in $A$.

represents a pseudo-formula and $t$ a term, is used to denote the result of substituting $t$ for all free occurrences of $x$ in $A$.

Notations like $A_{xt}^{ax}$ are used to denote the result of first substituting $x$ for $a$ and then $t$ for $x$. To denote the result of carrying out these two substitutions simultaneously, I use the notation $\mathrm{S}_{xt}^{ax}A$.

In the sequel, I shall usually simply speak about formulas and related notions, tacitly understanding a reference to a language L. The letters $A$, $B$, $C$, $D$, $E$, and $F$ when standing alone will always represent formulas unless otherwise stated. As a part of another notation, such as $\forall xA$, they will also represent pseudo-formulas, but then the whole notation will always represent a formula.

The letters $\Gamma$ and $\Delta$ will represent sets of formulas.

The letters $t$ and $u$ will represent individual terms unless otherwise stated.

I shall at no occasion write down the symbols mentioned above, but shall only speak about them, using the syntactical notations indicated in the preceding paragraphs.[1] How the symbols actually look, we may thus leave indeterminate.

DEFINED SIGNS. There are no symbols for negation and equivalence in the languages L but "$\sim$" and "$\equiv$" are used syntactically to abbreviate names of formulas. Thus, $\sim A$ is defined to be the formula $A \supset \wedge$ and $A \equiv B$ to be the formula $(A \supset B) \,\&\, (B \supset A)$.

SUBFORMULAS. The notion of *subformula* is defined inductively by:

1) $A$ is a subformula of $A$.

2) If $B \,\&\, C$, $B \vee C$, or $B \supset C$ is a subformula of $A$, then so are $B$ and $C$.

3) If $\forall xB$ or $\exists xB$ is a subformula of $A$, then so is $B_t^x$.

## § 2. Inference rules and deductions

### A. INFORMAL ACCOUNT

PRELIMINARY EXPLANATIONS. I first give an informal account of how the deductions are obtained in the systems under consideration. All the notions that are introduced informally in this account will later be defined in part B of this section and in § 3.

---

[1] On the syntactical level, I use signs as names of themselves if there is no risk of confusion.

We start the deduction by inferring a consequence from some assumptions by means of one of the inference rules listed below. We indicate this by writing the formulas assumed on a horizontal line and the formula inferred immediately below this line. From the formula inferred—possibly together with some other formulas obtained in a similar manner or some other assumptions that we want to make—we now infer a new consequence in accordance with one of the inference rules. We again indicate this by arranging all the formulas involved in such a way that the premisses in the last inference come on a horizontal line and the consequence comes immediately below this line. Continuing in this way, we obtain successively larger configurations that have the form of a tree as in the example below, where we want to derive $A \supset (B \mathbin{\&} C)$ from $(A \supset B) \mathbin{\&} (A \supset C)$:

$$
\cfrac{
  \cfrac{A \qquad \cfrac{(A \supset B) \mathbin{\&} (A \supset C)}{A \supset B}}{B}
  \qquad\qquad
  \cfrac{A \qquad \cfrac{(A \supset B) \mathbin{\&} (A \supset C)}{A \supset C}}{C}
}{B \mathbin{\&} C}
$$

We note that, if a formula is used as a premiss in two different inferences, then it also occurs twice in the configuration.

At each step so far, the configuration is a deduction of the undermost formula from the set of formulas that occur as assumptions. The assumptions are the uppermost formula occurrences, and we say that the undermost formula depends on these assumptions. Thus, the example above is a deduction of $B \mathbin{\&} C$ from the set $\{(A \supset B) \mathbin{\&} (A \supset C), A\}$, and, in this deduction, $B \mathbin{\&} C$ is said to depend on the top-occurrences of these two formulas.

As a result of certain inferences, however, the formula inferred becomes independent of some or all assumptions, and we then say that we discharge the assumptions in question. There will be four ways to discharge assumptions, namely:

1) given a deduction of $B$ from $\{A\} \cup \Gamma$, we may infer $A \supset B$ and discharge the assumptions of the form $A$;

2) given a deduction of $\wedge$ from $\{\sim A\} \cup \Gamma$, we may infer $A$ and discharge the assumptions of the form $\sim A$;

3) given three deductions, one of $A \vee B$, one of $C$ from $\{A\} \cup \Gamma_1$, and one of $C$ from $\{B\} \cup \Gamma_2$, we may infer $C$ and discharge the assumptions of the form $A$ and $B$ that occur in the second and third deduction

respectively, i.e., below the end-formulas of the three deductions, we write $C$ and then obtain a new deduction of $C$ that is independent of the mentioned assumptions;

4) given one deduction of $\exists xA$ and one of $B$ from $\{A_a^x\} \cup \Gamma$, we may infer $B$ and discharge assumptions of the form $A_a^x$, provided that $a$ does not occur in $\exists xA$, in $B$, or in any assumption—other than those of the form $A_a^x$—on which $B$ depends in the given deduction.

To facilitate the reading of a deduction, one may mark the assumptions that are discharged by numerals and write the same numeral at the inference by which the assumption is discharged (see e.g. p. 21).

To continue the deduction in the example above, we may write $A \supset (B \mathbin{\&} C)$ below $B \mathbin{\&} C$ and obtain then a deduction of $A \supset (B \mathbin{\&} C)$ from $\{(A \supset B) \mathbin{\&} (A \supset C)\}$.

An example. Before I list the inference rules, I give one further example of a deduction and illustrate how it may correspond to informal reasoning. An informal derivation of $\forall x \exists y(Pxy \mathbin{\&} Pyx)$ from the two assumptions

(1)                    $\forall x \exists y Pxy$

(2)                    $\forall x \forall y(Pxy \supset Pyx)$

may run somewhat as follows:

From (1), we obtain

(3)                    $\exists y Pay.$

Let us therefore assume

(4)                    $Pab.$

From (2), we have

(5)                    $Pab \supset Pba,$

and from (4) and (5)

(6)                    $Pba.$

Hence, from (4) and (6), we obtain

(7)                    $Pab \mathbin{\&} Pba$

and from (7) we get

$$(8) \qquad\qquad \exists y(Pay \mathbin{\&} Pya)$$

Now, (8) is obtained from assumption (4), but the argument is independent of the particular value of the parameter $b$ that satisfies (4). In view of (3), we therefore have:

$$(9) \qquad\qquad \text{(8) is independent of the assumption (4).}$$

Because of (9), (8) depends only on (1) and (2) and thus holds on these assumptions for any arbitrary value of $a$. Hence, the desired result:

$$(10) \qquad\qquad \forall x \exists y(Pxy \mathbin{\&} Pyx).$$

The corresponding natural deduction is given below; the numerals refer to the steps in the informal argument above (rather than to the way the assumptions are discharged).

$$
\begin{array}{c}
\dfrac{\forall x \forall y(Pxy \supset Pyx)\ {}^{(2)}}{\forall y(Pay \supset Pya)} \\[4pt]
{}^{(4)}\ Pab \qquad \dfrac{}{Pab \supset Pba}\ {}^{(5)} \\[2pt]
{}^{(4)}\ Pab \qquad \dfrac{}{Pba}\ {}^{(6)} \\[2pt]
{}^{(1)}\ \forall x \exists y Pxy \qquad \dfrac{}{Pab \mathbin{\&} Pba}\ {}^{(7)} \\[2pt]
{}^{(3)}\ \dfrac{}{\exists y Pay} \qquad \dfrac{}{\exists y(Pay \mathbin{\&} Pya)}\ {}^{(8)} \\[2pt]
\dfrac{}{\exists y(Pay \mathbin{\&} Pya)}\ {}^{(9)} \\[2pt]
\forall x \exists y(Pxy \mathbin{\&} Pyx)\ {}^{(10)}
\end{array}
$$

INFERENCE RULES. The inference rules consist of an introduction (I) rule and an elimination (E) rule for each logical constant except $\wedge$. We have thus a &I-rule, a &E-rule, a ∨I-rule etc. In addition, there are two rules for the sentential constant $\wedge$, one used in intuitionistic logic called the $\wedge_i$-rule and one used in classical logic called the $\wedge_c$-rule. The rules are indicated by the figures below. Letters within parentheses indicate that the inference rule discharges assumptions as explained above.

&I)   $\dfrac{A \quad B}{A \& B}$        &E)   $\dfrac{A \& B}{A} \quad \dfrac{A \& B}{B}$

∨I)   $\dfrac{A}{A \vee B} \quad \dfrac{B}{A \vee B}$      ∨E)   $\dfrac{A \vee B \quad \overset{(A)}{C} \quad \overset{(B)}{C}}{C}$

⊃I)   $\dfrac{\overset{(A)}{B}}{A \supset B}$        ⊃E)   $\dfrac{A \quad A \supset B}{B}$

∀I)   $\dfrac{A}{\forall x A^{a}_{x}}$        ∀E)   $\dfrac{\forall x A}{A^{x}_{t}}$

∃I)   $\dfrac{A^{x}_{t}}{\exists x A}$        ∃E)   $\dfrac{\exists x A \quad \overset{(A^{x}_{a})}{B}}{B}$

$\curlywedge_{\text{I}}$)   $\dfrac{\curlywedge}{A}$        $\curlywedge_{\text{c}}$)   $\dfrac{\overset{(\sim A)}{\curlywedge}}{A}$

*Restriction on the* ∀I-*rule*: $a$ must not occur in any assumption on which $A$ depends.

*Restriction on the* ∃E-*rule*: $a$ must not occur in $\exists x A$, in $B$, or in any assumption on which the upper occurrence of $B$ depends other than $A^{x}_{a}$.

*Restriction on the* $\curlywedge$-*rules*: $A$ is to be different from $\curlywedge$. This restriction is, of course, not essential, but saves us from considering certain trivial cases in the sequel.

*Restriction on the* $\curlywedge_{c}$-*rule*: $A$ is not to have the form $B \supset \curlywedge$. This restriction is also only a matter of convenience. That nothing is lost by this restriction can easily be seen. Suppose we have an application of the $\curlywedge_{c}$-rule that does not satisfy the restriction. Then the assumption $(B \supset \curlywedge) \supset \curlywedge$ discharged by this application can be replaced with the following deduction of $(B \supset \curlywedge) \supset \curlywedge$ from $\{B\}$:

$$\begin{array}{c} \text{(1)} \\ \dfrac{B \qquad B \supset \curlywedge}{\dfrac{\curlywedge}{(B \supset \curlywedge) \supset \curlywedge}} \text{ (1)} \end{array}$$

By this replacement, the given application of the $\curlywedge_c$-rule is turned into an application of the $\supset$ I-rule discharging the assumption $B$.

REMARK 1. Gentzen stated also two rules for negation, namely:

$$\sim\text{I)} \quad \dfrac{\overset{(A)}{\curlywedge}}{\sim A} \qquad \sim\text{E)} \quad \dfrac{A \quad \sim A}{\curlywedge}$$

When we define negation as above, these rules become special cases of the $\supset$ I-rule and the $\supset$ E-rule respectively.

CLASSICAL, INTUITIONISTIC, AND MINIMAL LOGIC. The differences between classical, intuitionistic, and minimal logic have a simple and illuminating characterization in the systems of natural deduction. For deductions in *minimal logic*, we may use only the introduction and elimination rules for sentential connectives and quantifiers. For deductions in *intuitionistic logic*, we may use all these rules and in addition the $\curlywedge_i$-rule. For deductions in *classical logic*, we may use all the inference rules including the $\curlywedge_c$-rule (note that the $\curlywedge_i$-rule is a special case of the $\curlywedge_c$-rule).

REMARK 2. To get classical logic from intuitionistic logic, we can, instead of adding the inference rule $\curlywedge_c$, add all formulas of the form $A \vee \sim A$ as axioms. This is the course followed by Gentzen, who also considers the following inference rule for the elimination of double negation:

$$\dfrac{\sim \sim A}{A}$$

(By this addition, the $\curlywedge_i$-rule becomes redundant.)

This concludes the informal explanation of the Gentzen-type systems of natural deduction. I shall now give a more precise definition of the notion of a deduction and some related notions. (For another variant of Gentzen's systems, see Appendix C, § 2.)

## B. Definitions

FORMULA-TREES. As explained above, the deductions are certain trees of formulas, or shorter, *formula-trees*. If $\Pi_1$, $\Pi_2$, ..., $\Pi_n$ is a sequence[1] of formula-trees, then

(1)                     $(\Pi_1, \Pi_2, ..., \Pi_n/A)$

is to be the tree obtained by arranging the configurations $\Pi$'s and $A$ so that the $\Pi$'s end on a horizontal line immediately above $A$.[2] I shall write (1) also in the more graphical notation:

$$\frac{\Pi_1 \; \Pi_2 \; ... \; \Pi_n}{A}$$

NOTIONS CONCERNING INFERENCE RULES. The inference rules were indicated by certain figures above. I shall say that $(A_1, A_2, ..., A_n/A)$ is an *instance of a certain inference rule* if it has the form indicated by the corresponding figure. Thus, $(A_1, A_2, ..., A_n/B)$ is said to be an *instance of*:

the & *I-rule* if $n = 2$ and $B$ is $A_1 \& A_2$;

the &E-*rule* if $n = 1$ and $A_1$ is $(B \& C)$ or $(C \& B)$ for some $C$;

the vI-*rule* if $n = 1$ and $B$ is $(A_1 \vee C)$ or $(C \vee A_1)$ for some $C$;

the vE-*rule* if $n = 3$, $A_1$ has the form $(C \vee D)$ and $A_2 = A_3 = B$; etc.

If $(A_1, A_2, ..., A_n/B)$ is an instance of an inference rule, we call $A_1, A_2, ..., A_n$ the *premisses* and $B$ the *consequence* of this instance. In an instance $(A, A \supset B/B)$ of the $\supset$ E-rule, an instance $(B \vee C, A, A/A)$ of the vE-rule, or an instance $(\exists x B, A/A)$ of the $\exists$ E-rule, $A$ is said to be *minor premiss*. The first [second] occurrence of $A$ in the instance of the vE-rule is the first [second] minor premiss of this instance. A premiss that is not minor is a *major premiss*.

DEDUCTION RULES. The inference rules do not characterize a system of natural deduction completely, since it is not stated in them how assumptions are discharged, and since the use of certain inference

---

[1] By a sequence, I understand simply a list of elements (rather than a function as in the set-theoretical sense). In particular, a sequence of one term is identified with the term itself, and I shall say that $\alpha_i (i \leqslant n)$ belongs to or is an element in the sequence $\alpha_1, \alpha_2, ..., \alpha_n$. Empty sequences are not considered except when explicitly admitted.

[2] We have: $\Pi$ is a formula-tree if and only if either (i) $\Pi$ is a formula or (ii) $\Pi$ is $(\Pi_1, \Pi_2, ...., \Pi_n/A)$, where $\Pi_1, \Pi_2, ...,$ and $\Pi_n$ are formula-trees. Abstractly, we may understand (1) as just the ordered $(n + 1)$ -tuple $<\Pi_1, \Pi_2, ...., \Pi_n, A>$, but I shall usually have the more concrete interpretation in mind.

rules are circumscribed by restrictions which are formulated in terms of what assumptions the premisses depend on. In particular, to characterize the roles of $\vee$E, $\supset$I, $\forall$I, $\exists$E, and the principle for indirect proof, I shall state a number of rules that I shall call *deduction rules*. The inference rules stated above for $\vee$E, $\supset$I, $\forall$I, $\exists$E, and $\wedge_c$ will be called *improper inference rules* while the others will be called *proper inference rules*.

A deduction rule can be thought of as a rule that allows us to infer a formula from a whole deduction given as a "premiss" and to determine what set of formulas the inferred formula depends on. For instance, the deduction rule for $\vee$E will allow us to infer $C$ given three "premisses" consisting of a deduction of $A \vee B$ from $\Gamma_1$, a deduction of $C$ from $\Gamma_2$, and a deduction of $C$ from $\Gamma_3$. Further, it allows us to conclude that the inferred formula depends on the union of $\Gamma_1$, $\Gamma_2 - \{A\}$, and $\Gamma_3 - \{B\}$. When applying a deduction rule in the systems considered here, however, the only things we need to know about a given deduction of $A$ from $\Gamma$ are the $\Gamma$ and $A$. A deduction rule can therefore be taken as a relation between pairs of the form $\langle \Gamma, A \rangle$. The form of an *instance of* the different *deduction rules* is to be as specified below:

$\vee$E: $\langle\langle \Gamma_1, A \vee B \rangle, \langle \Gamma_2, C \rangle, \langle \Gamma_3, C \rangle, \langle \Delta, C \rangle\rangle$ where $\Delta = \Gamma_1 \cup (\Gamma_2 - \{A\}) \cup (\Gamma_3 - \{B\})$.

$\supset$I: $\langle\langle \Gamma, B \rangle, \langle \Delta, A \supset B \rangle\rangle$ where $\Delta = \Gamma - \{A\}$.

$\forall$I: $\langle\langle \Gamma, A \rangle, \langle \Gamma, \forall x A_x^a \rangle\rangle$ where $a$ does not occur in any formula of $\Gamma$.

$\exists$E: $\langle\langle \Gamma_1, \exists x A \rangle, \langle \Gamma_2, B \rangle, \langle \Delta, B \rangle\rangle$ where $\Delta = \Gamma_1 \cup (\Gamma_2 - \Gamma_3)$, $\Gamma_3$ being the set of all formulas $A_a^x$ such that $a$ does not occur in $\exists x A$, in $B$, or in any formula of $\Gamma_2$ except $A_a^x$.

$\wedge_c$: $\langle\langle \Gamma, \wedge \rangle, \langle \Delta, A \rangle\rangle$ where $\Delta = \Gamma - \{\sim A\}$ and $A$ is different from $\wedge$ and is not of the form $\sim B$.

SYSTEMS OF NATURAL DEDUCTION. A system of natural deduction of the kind considered here is determined by a set of proper inference rules, a set of deduction rules, and a set of axioms.[1] We consider three

---

[1] In a logical calculus, as this notion is defined by Church [2] e.g., the deductions are determined by a set of inference rules and a set of axioms. Deduction rules of the kind considered above have, however, the effective character that is essential in the idea of a logical calculus.

such systems in this chapter, all without axioms. The *system for minimal logic* contains exactly all the rules for introduction and elimination of the sentential connectives and quantifiers as described above; i.e., the proper inference rules &I, &E, $\vee$I, $\supset$E, $\forall$E, and $\exists$I and the deduction rules $\vee$E, $\supset$I, $\forall$I, and $\exists$E. The *system for intuitionistic [classical] logic* contains all these rules and in addition the proper inference rule $\wedge_i$ [the deduction rule $\wedge_c$]. The systems for classical, intuionistic and minimal logic will be denoted by the letters C, I, and M. We get a system of natural deduction for a particular axiomatic theory that is expressed within a first order language by adding the set of axioms of the theory to one of these three systems.

DEDUCTIONS. For systems of the general kind described above, we now single out certain formula-trees as deductions. We first define what is meant by $\Pi$ being a *deduction in* a system S *of* a formula $A$ *depending on* a set $\Gamma$ of formulas:

1) If $A$ is not an axiom in S, then $A$ is a deduction in S of $A$ depending on $\{A\}$.

2) If $A$ is an axiom in S, then $A$ is a deduction in S of $A$ depending on the empty set.

3) If $\Pi_i$ is a deduction in S of $A_i$ depending on $\Gamma_i$ for every $i \leqslant n$, then $(\Pi_1, \Pi_2, ..., \Pi_n/B)$ is a deduction in S of $B$ depending on $\Delta$ provided that either

(i) $(A_1, A_2, ..., A_n/B)$ is a proper inference rule in S and $\Delta$ is the union of all $\Gamma_i$ for $i \leqslant n$, or

(ii) $\langle\langle \Gamma_1, A_1\rangle, \langle \Gamma_2, A_2\rangle ..., \langle \Gamma_n, A_n\rangle, \langle \Delta, B\rangle\rangle$ is an instance of a deduction rule in S.

$\Pi$ is a *deduction* in the system S *of A from* $\Gamma$ if and only if $\Pi$ is a deduction in S of $A$ depending on $\Gamma$ or some subset of $\Gamma$.

$A$ is *deducible from* $\Gamma$ in the system S—abbreviated as $\Gamma \vdash_s A$—if and only if there is a deduction in S of $A$ from $\Gamma$.[1]

A deduction of $A$ depending on the empty set is a *proof* of $A$; $A$ is *provable* in the system S—abbreviated as $\vdash_s A$—if and only if there is a proof in S of $A$.

---

[1] When $\Gamma$ is a unit set $\{B\}$, I shall sometimes simply say that $\Pi$ is a deduction of $A$ from $B$ and shall write: $B \vdash_s A$. Sometimes I simply speak about deductions, tacitly understanding a reference to a system S.

# § 3. Notions concerning deductions

To be able to speak about deductions conveniently, I need several concepts, which I bring together in this section; most of them are probably clear from their wording or from the informal explanations in the first part of § 2.

Notions concerning trees. The following notions, here stated for formula-trees, are to apply also to trees formed by elements other than formulas.

The letter $\Pi$ will always stand for trees, and the letter $\Sigma$ for sequences of trees including the empty one. If $\Sigma$ is empty, we define $(\Sigma/A)$ to be equal to $A$.

I take for granted the notion of an *occurrence of a formula* or (synonymously) a *formula occurrence* in a formula-tree. Two formula occurrences are said to be of the same form or shape if they are occurrences of the same formula; they are identical only if they also stand at the same place in the formula-tree. I use the letters $A$, $B$, ..., $F$ [$\Gamma$ and $\Delta$] to stand for [sets of] formula occurrences as well.[1] Sometimes, when I want to distinguish between two occurrences of a formula $A$, I use superscript notation, like $A^1$, $A^2$, etc.

I also take for granted the meaning of saying that a formula occurrence $A$ stands *immediately above* a formula occurrence $B$ (or that $B$ stands *immediately below* $A$) in a formula-tree $\Pi$.

A *top-formula* in a formula-tree $\Pi$ is a formula-occurrence that does not stand immediately below any formula occurrence in $\Pi$. The *end-formula* of $\Pi$ is the formula occurrence in $\Pi$ that does not stand immediately above any formula occurrence in $\Pi$.

A sequence $A_1$, $A_2$, ..., $A_n$ of formula occurrences in a formula-tree $\Pi$ is a *thread* in $\Pi$ if (1) $A_1$ is a top-formula in $\Pi$, (2) $A_i$ stands immediately above $A_{i+1}$ in $\Pi$ for each $i < n$, and (3) $A_n$ is the end-formula of $\Pi$. When that is the case, $A_i$ is said to stand *above* [*below*] $A_j$ if $i < j$ [$i > j$].

If $A$ is a formula occurrence in the tree $\Pi$, the *subtree of $\Pi$ determined by $A$* is the tree obtained from $\Pi$ by removing all formula occurrences except $A$ and the ones above $A$.

Let $A$ be a formula occurrence in $\Pi$, let $(\Pi_1, \Pi_2, ..., \Pi_n/A)$ be the

---

[1] Also in the same context, the letters $A$, $B$, ... may be used to represent both a certain formula and a particular occurrence of this formula, but in the text, it will then usually be stated in each case whether the shape or the occurrence is intended.

subtree of $\Pi$ determined by $A$, and let $A_1, A_2, ..., A_n$ be the end-formulas of $\Pi_1, \Pi_2, ..., \Pi_n$ respectively. We then say that $A_1, A_2, ..., A_n$ are the formula occurrences immediately above $A$ in $\Pi$ in *their order from left to right*. We shall then also say that $A_i$ is *side-connected* with $A_j$ $(i, j \leqslant n)$.

If $\Gamma$ is a set of top-formulas in $\Pi$, then $(\Sigma/\Gamma/\Pi)$ is to be the tree obtained as the result of writing $\Sigma$ above each top-formula in $\Pi$ that belongs to $\Gamma$ (i.e. so that the end-formulas of the trees in the sequence $\Sigma$ come on a horizontal line immediately above each such top-formula).[1] If $\Sigma$ or $\Gamma$ is empty, $(\Sigma/\Gamma/\Pi) = \Pi$.

When $\Gamma$ is the unit set of a formula occurrence $A$ in $\Pi$, I will write simply $(\Sigma/A/\Pi)$; and, when $\Gamma$ is a set of formula occurrences of the shape $A$, I will sometimes denote this set by $[A]$ and write $(\Sigma/[A]/\Pi)$. In a more graphic notation, I write in the respective cases:

$$\frac{\Sigma}{(A)} \qquad \frac{\Sigma}{[A]}$$
$$\Pi \qquad\qquad \Pi$$

The *length* of a formula-tree is the number of formula occurrences in the treee.

The notation used for substitution in formulas is also used in connection with formula-trees and sequences of such trees. Thus, $\Sigma_t^a$ denotes the result of carrying out the substitution in question in all formulas that occur in the trees that belong to the sequence $\Sigma$.

APPLICATIONS OF INFERENCE RULES. Let $B$ be a formula occurrence in a deduction $\Pi$ and let $A_1, A_2, ..., A_n$ be all the formula occurrences immediately above $B$ in $\Pi$ in their order from left to right. Then $\alpha = (A_1, A_2, ..., A_n/B)$ has the form of an instance of an inference rule R and I shall say that $\alpha$ is an *application of* R *in* $\Pi$. It can happen that $\alpha$ has the form of an instance of both the $\wedge$-rule and some other rule. I shall then follow the convention of considering $\alpha$ to be an application of only the $\wedge$-rule. It then holds that $\alpha$ is an application of at most one inference rule. In the case above, I shall say that $A_i$ $(i \leqslant n)$ is a *premiss*

---

[1] More precisely, we may define this operation by the following recursion, where it is assumed that $\Gamma$ is a non-empty set of formula occurrences in $\Pi$.

1) If $\Pi = A$, then $(\Sigma/\Gamma/\Pi) = (\Sigma/A)$.

2) If $\Pi = (\Pi_1, \Pi_2, ..., \Pi_n/A)$ and $\Gamma_i$ $(1 \leqslant i \leqslant n)$ are the members of $\Gamma$ that belong to $\Pi_i$, then $(\Sigma/\Gamma/\Pi) = ((\Sigma/\Gamma_1/\Pi_1), (\Sigma/\Gamma_2/\Pi_2), ..., (\Sigma/\Gamma_n/\Pi_n)/A)$.

and that $B$ is the *consequence of the application* $\alpha$ *of* $R$ (sometimes leaving out "the application of" for brevity).

ASSUMPTIONS. Let $\Pi$ be a deduction in a system $S$ and let $A$ be a top-formula in $\Pi$. Then, the formula occurrence $A$ is said to be an *axiom in* $\Pi$ if it has the form of an axiom in $S$ and is said to be an *assumption in* $\Pi$ if it does not have such a form.

A formula occurrence $A$ in a deduction $\Pi$ is said to *depend* in $\Pi$ *on the set* $\Gamma$ *of formulas* if the subtree of $\Pi$ determined by $A$ is a deduction depending on $\Gamma$. If $B \in \Gamma$, then we say that $A$ *depends* in $\Pi$ *on the formula* $B$. We sometimes need to consider also the assumptions, i.e., the particular top-formulas, on which $A$ depends. For the system for classical logic, we define:

Let $A$ be an assumption in a deduction $\Pi$ and let $\tau$ be the thread that begins with $A$. Then, $A$ is *discharged in* $\Pi$ *at* $B$ *by the application* $\alpha$ *of* $R$ if and only if $B$ is the first formula occurrence $C$ in $\tau$ such that one of the following four conditions holds:

1) R is $\vee$E, the major premiss of $\alpha$ has the form $A \vee D$ or $D \vee A$ (for some $D$), and $C$ is the first or second minor premiss of $\alpha$ respectively.

2) R is $\supset$I, $C$ is the premiss of $\alpha$, and the consequence of $\alpha$ has the form $A \supset C$.

3) R is $\exists$E, the major premiss of $\alpha$ has the form $\exists x A_x^a$, $C$ is the minor premiss of $\alpha$, and $a$ does not occur in $C$ or in any formula—other than the one of the form $A$—on which $C$ depends.

4) R is $\lambda_c$, $C$ is the premiss of $\alpha$, $A$ has the form $\sim D$, where $D$ in its turn is not a negation, and the consequence of $\alpha$ is $D$.

For the systems for intuitionistic and minimal logic we make the same definition but leave out clause 4).

A formula occurrence $B$ is said to *depend* in the deduction $\Pi$ *on the assumption* $A$ if $B$ belongs to the thread $\tau$ in $\Pi$ that begins with $A$ and $A$ is not discharged in $\Pi$ at a formula occurrence that precedes $B$ in $\tau$.

PROPER PARAMETERS. A parameter $a$ is said to be the *proper parameter of an application* $\alpha$ *of* $\forall$I in $\Pi$ if $\alpha$ has the form $(A/\forall x A_x^a)$ and $a$ actually occurs in $A$. A parameter $a$ is said to be the *proper parameter of an application* $\alpha$ *of* $\exists$E *in* $\Pi$ if $\alpha$ has the form $(\exists x A_x^a, B/B)$, an assumption of the form $A$ is discharged by $\alpha$, and $a$ actually occurs in $A$. A parameter $a$ is said to be a proper parameter in a deduction $\Pi$ if it is the proper parameter of some application of $\forall$I or $\exists$E in $\Pi$.

PURE PARAMETERS. In the rest of this section, I deal with certain details concerning proper parameters in order to facilitate operations on deductions that will be carried out in the sequel. In most cases, however, it is sufficient that the proper parameters are chosen so that they satisfy clauses (i)–(iii) in the lemma on parameters below. The reader may content himself with realizing that this is possible.

By a *connection* in a deduction $\Pi$ between two formula occurrences $A$ and $B$ in $\Pi$, I understand a sequence $A_1, A_2, ..., A_n$ of formula occurrences in $\Pi$ such that $A_1 = A$, $A_n = B$, and one of the following conditions holds for each $i < n$:

1) $A_i$ is not major premiss of an application of $\vee E$ and $\exists E$, and $A_{i+1}$ stands immediately below $A_i$; or vice versa;

2) $A_i$ is premiss of $\supset E$, and $A_{i+1}$ is side connected with $A_i$;

3) $A_i$ is major premiss of an application of $\vee E$ or $\exists E$, and $A_{i+1}$ is an assumption discharged by this application; or vice versa;

4) $A_i$ is a consequence of an application of $\supset I$ or the $\wedge_c$-rule, and $A_{i+1}$ is an assumption discharged by this application; or vice versa.

I shall say that a formula occurrence $B$ is *linked* in the deduction $\Pi$ *to* a formula occurrence $A$ *by* a parameter $a$ if there is a connection between $A$ and $B$ in $\Pi$ such that $a$ occurs in every formula occurence in the connection.

*Remark.* We may note in passing that the restrictions on the $\forall I$- and $\exists E$-rules can be liberalized. Thus, it is sufficient to require of an application $\alpha$ of the $\forall I$-rule that the premiss of $\alpha$ is not linked by the proper parameter of $\alpha$ to any assumption on which it depends; and to require of an application $\alpha$ of the $\exists E$-rule with major premiss $\exists x A$ and minor premiss $B$ that, if $A_a^x$ is an assumption discharged by $\alpha$, then $A_a^x$ is not linked by $a$ to $\exists x A$, to $B$, or to any assumption on which $B$ depends that differs in shape from that of $A_a^x$. If this more liberal restriction is adopted, the following is seen to hold: If $\Pi_1$ is a deduction of $A$ from $\Gamma$ and $\Pi_2$ is a deduction of B from $\{A\} \cup \Delta$ and $[A]$ is the set of assumptions of the form $A$ on which the end-formula of $\Pi_2$ depends, then $(\Pi_1/[A]/\Pi_2)$ is a deduction of $B$ from $\Gamma \cup \Delta$. — However, for our purpose, it will be more convenient to follow an opposite course, putting stronger restrictions on the proper parameters as follows.

I shall say that a proper parameter of an application $\alpha$ of $\forall I$ [$\exists E$] in a deduction $\Pi$ is *pure* in $\Pi$ if every formula occurrence in $\Pi$ in which $a$

occurs is linked by $a$ to the premiss of $\alpha$ [an assumption discharged by $\alpha$] and $a$ is the only proper parameter of $\alpha$ (the latter being trivial in case of $\forall$I). I shall say that a deduction has *only pure parameters* when every proper parameter in $\Pi$ is pure.

We can always transform a deduction $\Pi$ so that all its proper parameters become pure. In addition, we can chose the proper parameters so that they come to belong to a given set $K$ of individual parameters provided that $K$ is sufficiently large (e.g., infinite). Namely, we take an uppermost application $\alpha$ of $\forall$I or $\exists$E that has a proper parameter that is not pure or does not belong to $K$ (i.e., all applications of $\forall$I or $\exists$E with the consequence standing above the consequence of $\alpha$ are supposed to have proper parameters (if any) that are pure and belongs to $K$) and then we substitute a parameter $b$ that belongs to $K$ and does not occur in $\Pi$ for all occurrences of every proper parameter $a$ of $\alpha$ that belongs to formula occurrences linked by $a$ to (i) the premiss of $\alpha$ in case $\alpha$ is an application of $\forall$I, (ii) an assumption discharged by $\alpha$ in case $\alpha$ is an application of $\exists$E. Then $b$ is pure in the new deduction and pure parameters remain pure. We can obviously repeat this procedure until all proper parameters become pure and members of $K$, and we so have the following lemma. The three clauses (i)–(iii) are easy consequences of the fact that the parameters are pure.

LEMMA ON PARAMETERS. *If $\Gamma \vdash_s A$ and $K$ is an infinite set of individual parameters, then there is a deduction $\Pi$ in S of $A$ from $\Gamma$ such that all its proper parameters are pure and belong to $K$. In particular, $\Pi$ then satisfies the following three clauses:*

(i) *The proper parameter of an application $\alpha$ of $\forall I$ in $\Pi$ occurs in $\Pi$ only in formula occurrences above the consequence of $\alpha$.*

(ii) *The proper parameter of an appliation of $\exists E$ in $\Pi$ occurs in $\Pi$ only in formula occurrences above the minor premiss of $\alpha$.*

(iii) *Every proper parameter in $\Pi$ is a proper parameter of exactly one application of the $\forall I$-rule or the $\exists E$-rule in $\Pi$.*

## § 4. An alternative definition of the deductions

With the help of the notions defined in § 3, a slightly different notion of deduction could have been defined. This alternative definition will, however, not be used before Chapter VI; the present section may thus be postponed until then.

Let all the inference rules (i.e., also the improper ones) be given with a system S of natural deduction. Let us say that $\Pi$ is a *quasi-deduction* in a system S when $\Pi$ is a formula-tree such that, if $B$ is a formula occurrence in $\Pi$ and $A_1, A_2, ..., A_n$ are all the formula occurrences immediately above $B$ in $\Pi$ in their order from left to right, then $(A_1, A_2, ..., A_n/B)$ is an application of an inference rule in S.

Axioms and assumptions in a quasi-deduction are defined as for deductions.

By a *discharge-function* $\mathcal{J}$ *for a quasi-deduction* $\Pi$, we understand a function from a set of assumptions in $\Pi$ that assigns to $A$ either $A$ itself or a formula occurrence in $\Pi$ below $A$.

Let $\Pi$ be a quasi-deduction and let $\mathcal{J}$ be a discharge-function for $\Pi$. We say that an assumption $A$ in $\Pi$ is discharged with respect to $\mathcal{J}$ at $B$ if $\mathcal{J}(A) = B$. $B$ is said to depend with respect to $\mathcal{J}$ on the assumption $A$ if $B$ belongs to the thread $\tau$ in $\Pi$ that begins with $A$ and $A$ is not discharged with respect to $\mathcal{J}$ at a formula occurrence that precedes $B$ in $\tau$. $B$ is said to depend with respect to $\mathcal{J}$ on the formula $A$ if $B$ depends on an assumption of the shape $A$.

For classical [intuitionistic or minimal] logic, we define $\mathcal{J}$ as a *regular discharge-function* for a quasi-deduction $\Pi$ if (i) the proper parameter in an application of $\forall I$ does not occur in any assumption on which the premiss of this application depends and (ii) $\mathcal{J}(A)$ is a premiss $C$ in an application $\alpha$ of a rule R satisfying one of the conditions 1)-4) [1)-3)] in the definition on p. 27 of the phrase "$A$ is discharged in $\Pi$ at $B$ by the application $\alpha$ of R".

To say that $\Pi$ is a deduction of $A$ from $\Gamma$ as defined above is clearly equivalent to saying that $\Pi$ is a quasi-deduction such that the end-formula has the form $A$ and such that there exists a regular discharge-function $\mathcal{J}$ for $\Pi$ with the property that the end-formula of $\Pi$ depends with respect to $\mathcal{J}$ on formulas of $\Gamma$ only.

However, note that, by the first definition, an assumption is discharged as early as possible whereas it is not required of a regular discharge-function $\mathcal{J}$ that $\mathcal{J}(A)$ be the uppermost formula occurrence in the thread starting with $A$ that satisfies the stipulated conditions. In certain connections, it is an advantage not to require that an assumption be discharged as early as possible. We then use a *different notion of a deduction* defining a deduction of $A$ depending on $\Gamma$ as a pair $\langle \Pi, \mathcal{J} \rangle$, such that (i) $\Pi$ is a quasi-deduction with an end-formula of the shape $A$ and (ii) $\mathcal{J}$ is a regular discharge-function for $\Pi$ such that the end-formula of $\Pi$ depends with respect to $\mathcal{J}$ exactly on the formulas of $\Gamma$. This way of defining the deductions is more in agreement with the way of Gentzen. Usually, however, it is more convenient to let the dependency on the assumptions be uniquely given with the formula-tree as first defined.

## II. THE INVERSION PRINCIPLE

### § 1. Properties of the inference rules

The inference rules of the systems defined in Chapter I have some noteworthy properties. As we saw there, they comprise, with the exception of the $\wedge$-rules, exactly one introduction rule and one elimination rule for each sentential connective and quantifier.

INTRODUCTION RULES. An introduction rule for a logical constant $\gamma$ allows the inference *to* a formula $A$ that has $\gamma$ as principal sign from formulas that are subformulas of $A$; i.e. an instance of such a rule has a consequence $A$ with $\gamma$ as principal sign and has one or two premisses which are subformulas of $A$. The I-rule for $\gamma$ thus gives a sufficient condition for deducing formulas that have $\gamma$ as principal sign, which is stated in terms of subformulas of these formulas. We may note that this condition is closely connected with the meaning usually attributed to $\gamma$ when constructively interpreted,[1] and that it also agrees with the way formulas having $\gamma$ as principal sign are often derived in informal arguments. To make this more explicit, let us spell out these sufficient conditions that are given by the I-rules as follows.

| To deduce: | it is sufficient to have: |
|---|---|
| $A \& B$ (from $\Gamma$) | deductions both of $A$ and of $B$ (from $\Gamma$) |
| $A \vee B$ (from $\Gamma$) | a deduction either of $A$ or of $B$ (from $\Gamma$) |
| $A \supset B$ (from $\Gamma$) | a deduction of $B$ from $A$ (from $\Gamma \cup \{A\}$) |
| $\forall x A$ (from $\Gamma$) | a deduction of $A_a^x$ (from $\Gamma$), where $a$ is a parameter not in $\forall x A$ (nor in any formula of $\Gamma$) |
| $\exists x A$ (from $\Gamma$) | a deduction of $A_t^x$ (from $\Gamma$) for some term $t$ |

ELIMINATION RULES. An elimination rule for a logical constant $\gamma$ allows an inference *from* a formula that has $\gamma$ as principal sign, i.e. an instance of such a rule has a major premiss $A$, which has $\gamma$ as principal

---

[1] Cf., e.g., Heyting [1], pp. 97–99 and 102; Hilbert–Bernays [1], Ch. I; and Gentzen [4], Abschnitt III.

sign. In an application of such a rule, subformulas of the major premiss occur either as consequence and as minor premiss (if any) or as assumptions discharged by the application.

INVERSION PRINCIPLE. Observe that an elimination rule is, in a sense, the inverse of the corresponding introduction rule: by an application of an elimination rule one essentially only restores what had already been established if the major premiss of the application was inferred by an application of an introduction rule.[1] This relationship between the introduction rules and the elimination rules is roughly expressed by the following principle, which I shall call the *inversion principle*:[2]

*Let α be an application of an elimination rule that has B as consequence. Then, deductions that satisfy the sufficient condition (in the list above) for deriving the major premiss of α, when combined with deductions of the minor premisses of α (if any), already "contain" a deduction of B; the deduction of B is thus obtainable directly from the given deductions without the addition of α.*

I illustrate this principle by two examples (a more detailed verification follows in § 2):

1. $A$ may be inferred from $A \& B$ by $\&E$. In agreement with the inversion principle it holds: the above condition for deducing $A \& B$ from $\Gamma$ contains a deduction of $A$ from $\Gamma$.

2. $B$ may be inferred from $A$ and $A \supset B$ by $\supset E$. In agreement with the inversion principle it holds: the condition given above for proving $A \supset B$, i.e. a deduction of $B$ from $\{A\}$, combined with a proof of the minor premiss $A$ constitutes a proof of $B$.

The inversion principle says in effect that nothing is "gained" by inferring a formula through introduction for use as a major premiss in

---

[1] With Gentzen [3] (p. 189) one may say metaphorically that "an introduction rule gives, so to say, a definition of the constant in question", while "an elimination rule is only a consequence of the corresponding introduction rule, which may be expressed somewhat as follows: at an inference by an elimination rule, we are allowed to 'use' only what the principal sign of the major premiss 'means' according to the introduction rule for this sign".

[2] What Lorenzen has called the principle of inversion (see e.g. Lorenzen [1] and [2]) is closely related to this idea emanating from Gentzen. (For a correct statement of Lorenzen's principle, see Hermes [1].)

an elimination. The principle thus suggests the following *inversion theorem*:

> If $\Gamma \vdash A$, then there is a deduction of A from $\Gamma$ in which no formula occurrence is both the consequence of an application of an I-rule and major premiss of an application of an E-rule.

This theorem, with some slight additions, is fundamental for the sequel and gives a result for systems of natural deduction that is equivalent to Gentzen's Hauptsatz for the calculi of sequents. The central idea of Gentzen's proof of the Hauptsatz can be said also to be based on the inversion principle.[1] The proof of the theorem above, however, is somewhat more directly based on this principle and is strikingly simple; indeed, the proof is already suggested in the examples above.

We note that a consequence of an I-rule which is also major premiss of an E-rule constitutes a complication in a deduction. As such a complication can be removed according to the inversion theorem, we may ask whether it is possible to transform every deduction to a corresponding "normal" one which proceeds, so to say, directly, without any detours, from the assumptions to the end-formula. That such is the case will be shown in the next chapters by some slight additions to the inversion theorem. Some additions are necessary since we have to consider also the complications that can arise from applications of the $\lambda$-rules. A formula occurrence in a deduction $\Pi$ that is the consequence of an application of an I- or $\lambda$-rule and major premiss of an application of an E-rule is said to be a *maximum formula* in $\Pi$. In a natural sense, we can say that a maximum formula is of higher degree than "surrounding" formula occurrences (compare the definition of paths on p. 52).

REMARK. I conclude this section with a remark concerning negation in classical logic, which also gives an illustration of the inversion theorem in the form of a case in which the principle fails. The $\lambda$-rules clearly fall outside the division into I- and E-rules. One may consider the possibility of replacing the sentential constant $\lambda$ with a sentential connective for negation as primitive sign. In the case of classical logic,

---

[1] As is also remarked by Gentzen [5].

we can then replace the $\wedge_c$-rule with two $\sim$-rules, indicated by the following figures (cf. Remark 2 on p. 21):

$$\sim\text{I)} \quad \frac{\begin{array}{cc}(A) & (A)\\ B & \sim B\end{array}}{\sim A} \qquad \sim\text{E)} \quad \frac{\sim\sim A}{A}$$

The inference rule for $\sim$I is thus improper, and an instance of the corresponding deduction rule is to have the form

$$\langle\langle\Gamma_1, B\rangle, \langle\Gamma_2, \sim B\rangle, \langle\Delta, \sim A\rangle\rangle, \text{ where } \Delta = (\Gamma_1 - \{A\}) \cup (\Gamma_2 - \{A\}).$$

(Premises of instances of the $\sim$I- and $\sim$E-rules are considered to be major premisses.)

We note that the $\sim$I-rule above does not have the same character as the other I-rules. Furthermore, we note that the $\sim$-rules do not satisfy the inversion principle, since it does not hold in general that a deduction of $A$ is in any way contained in the sufficient condition given by the $\sim$I-rule for deducing $\sim\sim A$ (i.e. two deductions, one of $B$ from $\{\sim A\}$ and one of $\sim B$ from $\{\sim A\}$). To show that the inversion theorem fails for this system, we may consider a proof of $A \vee \sim A$. It can be shown that a proof of $A \vee \sim A$ in this system has to use a formula occurrence both as consequence of an application of an I-rule and as a major premiss of an application of an E-rule as in the following example, where $\sim\sim(A \vee \sim A)$ is so used:

$$\frac{\begin{array}{c}(1)\\ \dfrac{A}{A \vee \sim A} \quad \sim(A \vee \sim A)\ (2)\end{array}}{\begin{array}{c}\dfrac{\sim A}{A \vee \sim A} \quad \sim(A \vee \sim A)\ (2)\\ \dfrac{}{\sim\sim(A \vee \sim A)}\ (2)\end{array}}$$

$$\frac{\sim\sim(A \vee \sim A)}{A \vee \sim A}$$

## § 2. Reduction steps

In this section, I shall show how to remove a formula occurrence that is the consequence of an I-rule and the major premiss of an E-rule. This constitutes the induction step in the proof of the inversion theorem, which will be proved in some stronger forms in the next chapters.

Let $\Pi$ be a deduction of $E$ from $\Gamma$ that has only pure parameters and that contains a formula occurrence $F$ that is a consequence of an application of an I-rule and major premiss of an application of an E-rule. Then $\Pi'$ is said to be a *reduction of* $\Pi$ *at* $F$ if $\Pi'$ is obtained from $\Pi$ by removing $F$ in the way described below. We have five cases depending on which logical constant is the principal sign of $F$ (clearly $F$ cannot be atomic). If the principal sign of $F$ is $\gamma$, then $\Pi'$ is said to be a $\gamma$-reduction of $\Pi$. In each case, it is easily verified that also $\Pi'$ is a deduction of $E$ from $\Gamma$.

&-*reduction.* $F$ then has the form $A \& B$ and is consequence of an application of &I and premiss of an application of &E. Immediately above $F$, we thus have occurrences of $A$ and $B$, and immediately below $F$, we have an occurrence either of $A$ or of $B$, say $A$; the other case is quite symmetrical. Hence $\Pi$ has the form shown to the left below. $\Pi'$ is to be as shown to the right.

$$
\begin{array}{cc}
\dfrac{\Sigma_1}{A} \quad \dfrac{\Sigma_2}{B} & \\
\hline
A \& B & \dfrac{\Sigma_1}{(A)} \\
(A) & \Pi_3 \\
\Pi_3 &
\end{array}
$$

v-*reduction.* Then, $F$ has the form $A \vee B$, and $\Pi$ has either the form shown to the left below, or a similar form with the difference that the formula occurring immediately above $F$ is $B$ instead of $A$, which case is quite symmetrical to the first one. $[A]$ and $[B]$ are to be the sets of the assumptions of the form $A$ and $B$, respectively, that are discharged by the application of vE in which $F$ is major premiss, and $\Pi'$ is to have the form shown to the right below.

$$
\begin{array}{cccc}
\dfrac{\Sigma_1}{A} & \dfrac{[A]}{\quad} & \dfrac{[B]}{\quad} & \dfrac{\Sigma_1}{\quad} \\
 & \dfrac{\Sigma_2}{C} & \dfrac{\Sigma_3}{C} & [A] \\
\hline
A \vee B & C & C & \Sigma_2 \\
 & (C) & & (C) \\
 & \Pi_4 & & \Pi_4
\end{array}
$$

To see that $\Pi'$ is a deduction of $E$ from $\Gamma$, note that since $\Pi$ has only pure parameters, proper parameters of applications of $\forall$I and $\exists$E in $\Sigma_2$ do not occur in $\Sigma_1$ (cf. clauses 1 and 2 in the lemma on parameters, p. 29).

$\supset$-*reductions.* Then $F$ has the form $A \supset B$, and $\Pi$ has the form shown to the left below, where $[A]$ is to be the set of assumptions discharged by the application of $\supset I$ in which $F$ is the consequence. $\Pi'$ is to be the deduction shown to the right below.

$$
\begin{array}{cc}
 & [A] \\
 & \Sigma_2 \\
\dfrac{\Sigma_1}{A} \quad \dfrac{B}{A \supset B} \\
(B) \\
\Pi_3
\end{array}
\qquad
\begin{array}{c}
\Sigma_1 \\
[A] \\
\Sigma_2 \\
(B) \\
\Pi_3
\end{array}
$$

$\forall$-*reductions.* Then $F$ has the form $\forall x A_x^a$ and $\Pi$ has the form shown to the left below. As explained in Ch. I, $\Sigma_{1t}^a$ is the result of substituting $t$ for all occurrences of $a$ in $\Sigma_1$. $\Pi'$ is to be as shown to the right below.

$$
\begin{array}{c}
\Sigma_1 \\
\dfrac{A}{\forall x A_x^a} \\
(A_{xt}^{ax}) \\
\Pi_2
\end{array}
\qquad
\begin{array}{c}
\dfrac{\Sigma_{1t}^a}{(A_t^a)} \\
\Pi_2
\end{array}
$$

To verify that $\Pi'$ is still a deduction of $E$ from $\Gamma$ we note the following facts: (i) The proper parameters in the deduction $(\Sigma_1/A)$ are all different from both $a$ and $t$ (from $a$ by clause 3 and from $t$ by clauses 1 and 2 in the lemma on parameters). (ii) $A_{xt}^{ax} = A_t^a$, since there can be no free occurrence of $x$ in $A$. (iii) The assumptions in $\Sigma_1$ on which $F$ depends do not contain any occurrence of $a$ by the definition of the deduction rule for $\forall I$ and hence they are not affected by the substitution of $t$ for $a$ in $\Sigma_1$.

$\exists$-*reductions.* Then $F$ has the form $\exists x A$ and $\Pi$ has the form shown to the left below, where $[A_a^x]$ is to be the set of assumptions discharged by the application of $\exists E$ in question. By substituting $t$ for $a$ in $\Sigma_2$, the set $[A_a^x]$ goes over to a set $[A_t^x]$; note that $A_{at}^{xa} = A_t^x$, since $a$ does not occur in $A$, according to the definition of the deduction rule for $\exists E$. $\Pi'$ is to be the deduction shown to the right below.

$$
\begin{array}{ccc}
\Sigma_1 & [A^x_a] & \Sigma_1 \\
\dfrac{A^x_t}{\exists x A} & \dfrac{\Sigma_2}{B} & \dfrac{[A^x_t]}{\Sigma^a_{2t}} \\[1ex]
(B) & & (B) \\
\Pi_3 & & \Pi_3
\end{array}
$$

To verify that $\Pi'$ is still a deduction of $E$ from $\Gamma$ we note the following facts: (i) Proper parameters of applications of $\forall$I and $\exists$E in $\Sigma_2$ do not occur in $\Sigma_1$ and are all different from $a$ and $t$ (by clauses 1)–3) in the lemma on parameters). (ii) $a$ does not occur in $B$ by the definition of the deduction rule for $\exists$E. (iii) Assumptions in $\Sigma_2$ that the consequence of the application of $\exists$E in question depends on do not contain any occurrences of $a$ by the definition of the deduction rule for $\exists$E and are hence not affected by the substitution of $t$ for $a$ in $\Sigma_2$.

REMARK. When the deductions are defined in the alternative way described in § 4 of Chapter I, a reduction of a deduction $\langle \Pi, \mathcal{J} \rangle$ at a formula occurrence $F$ is to be the deduction $\langle \Pi', \mathcal{J}' \rangle$ where $\Pi'$ is the quasi-deduction obtained by a reduction of $\Pi$ at $F$ as described above and $\mathcal{J}'$ is like $\mathcal{J}$ except for the necessary and obvious modifications that are due to the fact that certain assumptions in $\Pi$ disappear and that $\Pi'$ may contain a number of subdeductions of the same form, where $\Pi$ contained only one subdeduction of this form (this will happen, e. g., in $\supset$-reductions if the set $[A]$ contains more than one element).

# III. NORMAL DEDUCTIONS IN CLASSICAL LOGIC

## § 1. Theorem on normal deductions

In this chapter, I consider the question of a normal form for deductions in classical logic. To minimize the disturbing effect of the $\wedge_c$-rule in this connection (cf. the Remark in Ch. II, § 1), I shall consider the system obtained from the system for classical logic defined in Chapter I by excluding the v- and ∃-rules. This smaller system will be denoted by C'.

The languages corresponding to C' are, accordingly, to contain only $\wedge$, &, ⊃, and ∀ as logical constants. Disjunction and existential quantification can be defined in one of the usual ways (e.g., $A \vee B$ can be taken to be the formula $\sim A \supset B$, and ∃$xA$ can be taken to be the formula $\sim \forall x \sim A$), and it can then easily be verified that the v- and ∃-rules hold as derived rules in C'. Hence, C' is still adequate for classical logic.

We can restrict applications of the $\wedge_c$-rule in C' to the case where the consequence is atomic, and we then easily prove that every deduction in C' can be transformed into a corresponding deduction containing no maximum formula. In the next section, it is shown that such a deduction is of a rather perspicuous form, and I shall say that a deduction in C' is *normal* if it contains no maximum formula.

THEOREM 1. *If* $\Gamma \vdash_{c'} A$, *then there is a deduction in* C' *of A from* $\Gamma$ *in which the consequence of every application of the* $\wedge_c$*-rule is atomic.*

*Proof.* Let $\Pi$ be a deduction in C' of $A$ from $\Gamma$ in which the highest degree of a consequence of an application of the $\wedge_c$-rule is $d$, where $d > 0$. Let $F$ be a consequence of an application $\alpha$ of the $\wedge_c$-rule in $\Pi$ such that its degree is $d$ but no consequence of an application of the $\wedge_c$-rule in $\Pi$ that stands above $F$ is of degree $d$. Then $\Pi$ has the form

$$[\sim F]$$
$$\Sigma$$
$$\frac{}{\curlywedge}$$
$$\frac{}{(F)}$$
$$\Pi_1$$

where $[\sim F]$ is the set of assumptions discharged by $\alpha$. $F$ has one of the shapes $B \& C$, $B \supset C$, and $\forall x B$. We remove this application of the $\curlywedge_c$-rule by transforming $\Pi$ in the respective cases to:

$$
\begin{array}{cccc}
(1) & (3) & (1)\quad(2) & (1) \\
B \& C \quad (2) & B \& C \quad (4) & B \quad B \supset C \quad (3) & \forall x A \quad (2) \\
\hline
B \quad \sim B & C \quad \sim C & C \quad \sim C & A^x_a \quad \sim A^x_a \\
\hline
\curlywedge & \curlywedge & \curlywedge & \curlywedge \\
[\sim(B \& C)] & [\sim(B \& C)] \;(3) & [\sim(B \supset C)] \;(2) & [\sim \forall x A] \;(1) \\
\Sigma & \Sigma & \Sigma & \Sigma \\
\curlywedge & \curlywedge & \curlywedge & \curlywedge \\
B & C \;(4) & C \;(3) & A^x_a \;(2) \\
\hline
(B \& C) & & (B \supset C) \;(1) & (\forall x A) \\
\Pi_1 & & \Pi_1 & \Pi_1
\end{array}
$$

Here, $a$ is to be a parameter that does not occur in $\Pi$. The new applications of the $\curlywedge_c$-rule that arise from this transformation have consequences of degrees less than $d$. Thus, by successively repeating the transformation, we finally obtain a deduction of $A$ from $\Gamma$ in which the consequence of every application of the $\curlywedge_c$-rule is atomic.

THEOREM 2. *If $\Gamma \vdash_{c} A$, then there is a normal deduction in $C'$ of $A$ from $\Gamma$.*

*Proof.* Let $\Pi$ be a deduction in $C'$ of $A$ from $\Gamma$ that has only pure parameters and that is as described in Theorem 1. Let $F$ be a maximum formula in $\Pi$ such that there is no other maximum formula in $\Pi$ of higher degree than that of $F$ and such that maximum formulas in $\Pi$ that stand above a formula occurrence side-connected with $F$ (if any) have lower degrees than $F$. Let $\Pi'$ be a reduction of $\Pi$ at $F$ (as defined in Ch. II, §2). The new maximum formulas that may arise from this reduction are all of lower degree than that of $F$. It is also easily seen that $\Pi'$ is still as described in Theorem 1. $\Pi'$ may have some proper parameters that are not pure but they can obviously be made pure again without changing the degree or the number of maximum formulas. Hence, by a finite number of similar reductions we obtain a normal deduction of $A$ from $\Gamma$.

## § 2. The form of normal deductions

The normal deductions have a rather perspicuous form. Consider a thread in a deduction $\Pi$ in $\mathsf{C}'$ such that no element of the thread is a minor premiss of $\supset \mathrm{E}$. We shall then see that if $\Pi$ is normal, the thread can be divided into two parts in the following way: There is one formula occurrence $A$ in the thread such that all formula occurrences of the thread above $A$ are premisses of applications of E-rules and all formula occurrences in the thread below $A$ (except the last one) are premisses of applications of I-rules. Thus, in the upper part of the thread (above $A$), we pass from the top-formula to simpler and simpler formula occurrences, which are subformulas of the ones that stand above, and in the lower part (below $A$), we pass to more and more complex formulas, which contain the ones that stand above as subformulas. $A$ is thus a minimum formula in the thread.

By a *branch* in a deduction, I shall understand an initial part $A_1, A_2, ..., A_n$ of a thread in the deduction such that $A_n$ is either (i) the first formula occurrence in the thread that is the minor premiss of an application of $\supset \mathrm{E}$ or (ii) the last formula occurrence of the thread (i.e. the end-formula of the deduction) if there is no such minor premiss in the thread. A branch that is also a thread and that thus contains no minor premiss of $\supset \mathrm{E}$ is called a *main branch*. What was said above holds for branches in general and is expressed in the following theorem:

THEOREM 3. *Let $\Pi$ be a normal deduction in $\mathsf{C}'$, and let $\beta = A_1, A_2, ..., A_n$ be a branch in $\Pi$. Then there is a formula occurrence $A_i$, called the minimum formula in $\beta$, which separates two (possibly empty) parts of $\beta$, called the* E-part *and the* I-part *of $\beta$, with the properties:*

1) *Each $A_j$ in the E-part (i.e., $j < i$) is a major premiss of an E-rule and contains $A_{j+1}$ as a subformula.*

2) *$A_i$, provided that $i \neq n$, is premiss of an I-rule or of the $\wedge_c$-rule.*

3) *Each $A_j$ in the I-part, except the last one, (i.e., $i < j < n$) is a premiss of an I-rule and is a subformula of $A_{j+1}$.*

*Proof.* It can be seen that the formula occurrences in $\beta$ that are major premisses of E-rules precede all formula occurrences in $\beta$ that are premisses of I-rules or the $\wedge_c$-rule. Otherwise, there would be a first formula occurrence in $\beta$ which is a major premiss of an E-rule but succeeds a premiss of an I-rule or the $\wedge_c$-rule, and such a formula occurrence would be a maximum formula, contrary to the assumption that $\Pi$ is normal.

Now, let $A_i$ be the first formula occurrence in $\beta$ that is premiss of an I-rule or the $\wedge_c$-rule or, if there is no such segment, let $A_i$ be $A_n$. It can then be seen that $A_i$ is a minimum formula as described in the theorem. Obviously, $A_i$ satisfies clauses 1) and 2). By what has been proved, every formula occurrence $A_j$ such that $i < j < n$ is a premiss of an I-rule or of the $\wedge_c$-rule. However, the latter possibility is excluded, since a premiss of the $\wedge_c$-rule is an occurrence of $\wedge$ and can be a consequence of an E-rule only (note the restriction on the $\wedge$-rules). Hence, clause 3) is also satisfied.

It will sometimes be useful to assign an *order* to the branches in a deduction $\Pi$ as follows: A main branch (i.e. a branch that ends with the end-formula of $\Pi$) has the order 0. A branch that ends with a minor premiss of an application of $\supset$E is assigned the order $n+1$ if the major premiss of this application belongs to a branch of order $n$.

COROLLARY 1. (Subformula principle.) *Every formula occurrence in a normal deduction in $C'$ of $A$ from $\Gamma$ has the shape of a subformula of $A$ or of some formula of $\Gamma$, except for assumptions discharged by applications of the $\wedge_c$-rule and for occurrences of $\wedge$ that stand immediately below such assumptions.*

Note that, if $\sim B$ is an assumption discharged by an application of the $\wedge_c$-rule in a normal deduction of $A$ from $\Gamma$, then according to Corollary 1, $B$ is a subformula of $A$ or of some formula of $\Gamma$.

*Proof.* Let $\Pi$ be a normal deduction in $C'$ of $A$ from $\Gamma$. We show that the corollary holds for all formula occurrences in a branch of order $p$ on the assumption that it holds for formula occurrences in branches of order less than $p$. (This is sufficient, since every formula occurrence in $\Pi$ belongs to some branch in $\Pi$.) Let $\beta$ be $A_1, A_2, ..., A_n$ and let $A_i$ be the minimum formula of $\beta$.

For $A_n$ the assertion is immediate (either $A_n = A$, or $A_n$ is a minor premiss of an application of $\supset$E with a major premiss of the form $A_n \supset B$ that belongs to a branch of order $p-1$). Hence, by Theorem 3, the corollary holds for all $A_j$ such that $i < j < n$.

If $A_1$ is not discharged by an application of the $\wedge_c$-rule, then either $A_1 \in \Gamma$ or $A_1$ is discharged by an application $\alpha$ of $\supset I$ such that the consequence of $\alpha$ has the form $A_1 \supset B$ and belongs to the I-part of $\beta$ or to some branch of order less than $p$. Hence, in this case, $A_1$ is a subformula of the described kind, and, by Theorem 3, the same holds for all $A_j$ such that $j \leqslant i$.

If $A_1$ is discharged by an application of the $\wedge_c$-rule, then either $A_1$ is a premiss of an I-rule and so $A_1 = A_i$, or $A_1$ is a major premiss of the $\supset E$-rule and so $A_2 = A_i =$ an occurrence of $\wedge$, or $A_1$ is a minor premiss of the $\supset E$-rule and so $A_1 = A_n$; hence, also in the latter three cases, the proof is complete.

It is possible, in fact, to say more about the manner in which different formula occurrences in a normal deduction of $A$ from $\Gamma$ are related to $A$ and the formulas of $\Gamma$. Let us define a formula to be a *positive* [*negative*] *subformula* of $A$ as follows:

1) $A$ is a positive subformula of $A$.

2) $B$ and $C$ are positive [negative] subformulas of $A$ when $B \& C$ is, and so is $B_t^x$ when $\forall x B$ is.

3) If $B \supset C$ is a positive [negative] subformula of $A$, then $B$ is a negative [positive] subformula of $A$ and $C$ is a positive [negative] subformula of $A$.

Thus, the definitions run like the definition of subformula except for clause 3) above by which the antecedent of an implication is not a subformula of the same kind as the implication itself. Let us say that $D$ is a *strictly positive subformula* of $E$ if $D$ is a subformula of $E$ and this fact follows from the definition of subformula without using the provision "if $B \supset C$ is a subformula of $A$, then so is $B$".

$B$ is said to be an *assumption-part* [*conclusion-part*] of the pair $\langle \Gamma, A \rangle$ if $B$ is a negative [positive] subformula of $A$ or a positive [negative] subformula of some formula of $\Gamma$. By a proof analogous to that of Corollary 1, we then have the following corollary:

COROLLARY 2. *Let* $\beta = A_1, A_2, ..., A_n$ *be a branch in a normal deduction in* **C'** *of* $A$ *from* $\Gamma$. *Then:*

1) *Either every formula occurring in the E-part of* $\beta$ *is an assumption-part of* $\langle \Gamma, A \rangle$ *and is a strictly positive subformula of* $A_1$, *or the E-part of* $\beta$ *consists exactly of an assumption discharged by an application of the* $\wedge_c$-*rule.*

2) *The minimum formula of $\beta$ is an assumption-part of $\langle \Gamma, A \rangle$ and is a strictly positive subformula of $A_1$ and, if different from $\wedge$, is also a conclusion-part of $\langle \Gamma, A \rangle$ and a strictly positive subformula of $A_n$.*

3) *Every formula occurring in the I-part of $\beta$ is a conclusion-part of $\langle \Gamma, A \rangle$ and is a strictly positive subformula of $A_n$.*

## § 3. Some further corollaries

As an illustration to the theorems above, I infer some well-known results as corollaries.

COROLLARY 3. C′ *is consistent; in particular,* $\wedge$ *is not provable in* C′.

*Proof.* Assume the contrary. Let $\Pi$ be a normal proof of $\wedge$, and let $\beta$ be a main branch in $\Pi$. By Theorem 3, the I-part of $\beta$ is empty. But the first formula occurrence in $\beta$ is then an assumption that is not discharged in $\Pi$, contrary to the assumption that $\Pi$ is a proof.[1]

Classical logic is to a large extent treated as a special case of intuitionistic logic in the present chapter.[2] We note that to every formula $A$ in one of the languages specified in Chapter I, there is a formula $A^*$, classically equivalent to $A$ (i.e. $\vdash_c A \equiv A^*$), that contains no occurrence of ∨ or ∃ and in which every occurrence of an atomic pseudo-formula different from $\wedge$ is negated, i.e. stands as antecedent in a part of $A^*$ of the form $(B \supset \wedge)$. We further have:

---

[1] The corollary can also be obtained directly from the subformula principle (Corollary 1), but the present proof is applicable also in cases where that principle fails.

[2] This is perhaps not the most natural procedure from a classical point of view, and one may therefore consider modifications of the Gentzen-type system for classical logic. One rather natural modification is to make the system more symmetrical with respect to & and ∨. In the present system, the deduction forks only upwards, and the forking is so to say conjunctively; one could now allow the deductions to fork also downwards, disjunctively, so that the deduction is allowed to fork from a disjunction $A \vee B$ into two branches, starting with $A$ and $B$ respectively. Such a system was presented by the author in a colloquium in Los Angeles in 1964. (A suggestion to such a system may be found in Kneale [1] but the rules are there stated without sufficient restrictions.) The present proof of the theorem on normal deductions can be carried over with some changes to that system, and we then also obtain Herbrand's theorem as a corollary.

COROLLARY 4. *If A and the formulas of $\Gamma$ contain no occurrences of $\vee$ or $\exists$ and if all occurrences of atomic pseudo-formulas different from $\wedge$ in these formulas are negated, then it holds: If $\Gamma \vdash_c A$, then $\Gamma \vdash_i A$.*[1]

*Proof.* Let $\Pi$ be a normal deduction in C′ of A from $\Gamma$ which has only pure parameters and in which the consequence of every application of the $\wedge_c$-rule is atomic. A consequence $B$ of an application of the $\wedge_c$-rule in $\Pi$ is then a minor premiss of an application of $\supset$ E. Otherwise, by Theorem 3, $B$ would be either the end-formula of $\Pi$, which is impossible because of the assumptions about $A$, or a premiss of an application $\alpha$ of an I-rule. In the latter case, the consequence of $\alpha$ would have the form $C \supset B$ or $\forall x B_x^a$ and, by Corollary 1, would be a subformula of $A$ or of some formula of $\Gamma$ contrary to the assumptions about $\Gamma$ and $A$ (note that $B \neq \wedge$ by the restriction on the $\wedge$-rules).

A deduction in C′ in which the applications of the $\wedge_c$-rule do not discharge any assumptions is also a deduction in I. Let now $B$ be the consequence of a particular application of the $\wedge_c$-rule in $\Pi$ by which (exactly) the assumptions in the non-empty set $[\sim B]$ are discharged. Then $\Pi$ has the form shown to the left below and can be transformed as shown to the right below.

$$
\begin{array}{cc}
\begin{array}{c}
[\sim B] \\
\Sigma_1 \\
\dfrac{\wedge \qquad \Sigma_2}{B \qquad \sim B} \\
(\wedge) \\
\Pi_3
\end{array}
&
\begin{array}{c}
\Sigma_2 \\
[\sim B] \\
\Sigma_1 \\
(\wedge) \\
\Pi_3
\end{array}
\end{array}
$$

---

[1] If the languages are modified so that they do not contain predicate parameters (only predicate constants), then one can show that in certain mathematical theories, the deduction rule for $\wedge_c$ holds also intuitionistically if restricted to applications where the consequence is atomic. This is the case in, e.g., elementary number theory (Peano arithmetic). Theorem 1 then immediately gives that every formula in Peano arithmetic that does not contain $\vee$ or $\exists$ is intuitionistically provable if classically provable. This result was first obtained by Gödel [1] and the main idea of Theorem 1 is thus due to him. (Gödel also requires that the formula does not contain $\supset$; a sign for negation is then taken as primitive. ~~The result for Peano arithmetic then also holds when predicate parameters are present.~~) The same results were found by Gentzen [2] (stated in a manuscript set in type but not published) and by Bernays (see Gentzen [4], p. 532). Corollary 4 as formulated above can be found in Gentzen [2] and in Kleene [1].

We can suppose that $B$ is chosen so that no application of the $\wedge_c$-rule in $\Sigma_2$ discharges any assumptions. By a number of similar transformations, we thus obtain a deduction in the system for intuitionistic logic of $A$ from $\Gamma$.

For the next corollary, let us say that a parameter or non-logical constant *occurs positively [negatively]* in a formula $A$ if it occurs in an occurrence of a pseudo-formula in $A$ which by substitution of parameters for free variables become a positive [negative] subformula of $A$.

COROLLARY 5. (Interpolation theorem.)[1] *If $\Gamma \vdash_{c'} A$, then there is a formula $F$, called an interpolation formula to $\Gamma$ and $A$, such that $\Gamma \vdash_{c'} F$ and $F \vdash_{c'} A$ and such that every parameter and non-logical constant that occur positively [negatively] in $F$ occur positively [negatively] both in $A$ and in some formula of $\Gamma$.*

We recall that the predicate symbols that occur in the minimum formula of a branch occur both in the first and the last formula occurrence in the branch. This suggests that an interpolation formula as described in the corollary can be constructed from minimum formulas using the logical constants. Which minimum formulas we are to choose and how they are to be combined are determined by an induction over the length of the deduction of $A$ from $\Gamma$, establishing the following lemma, which I state only for parameters:

LEMMA. *Let $\Pi$ be a normal deduction in $C'$ of $A$ depending on $\Gamma$, and let $\Gamma_1$ and $\Gamma_2$ be two disjoint sets such that $\Gamma_1 \cup \Gamma_2 = \Gamma$. Then there is a formula $F$, called an interpolation formula to $\langle \Gamma_1, \langle \Gamma_2, A \rangle \rangle$, such that $\Gamma_1 \vdash F$ and $\{F\} \cup \Gamma_2 \vdash A$ and such that every parameter that occurs positively [negatively] in $F$ occurs positively [negatively] in some formula of $\Gamma_1$ and negatively [positively] in some formula of $\Gamma_2 \cup \{\sim A\}$.*

*Proof of the lemma.* If $\Pi = A$, then $\Gamma = \{A\}$ and the lemma is trivial; because if $A \in \Gamma_1$, then $A$ is such an interpolation formula, and if $A \notin \Gamma_1$

---

[1] A theorem of this kind was first proved by Craig [1]. That the parameters occur in the same way (positively or negatively) in the interpolation formula as in the assumptions and the end-formula is an addition due to Lyndon [1]. The present proof is similar to a recent proof by Schütte [3] (which came to my attention after having finished this part of the study). For another recent proof (which is not constructive) see Henkin [2].

then $\wedge \supset \wedge$ is such an interpolation formula. The induction step is divided into three cases.

*Case I. The end-formula of $\Pi$ is the consequence of an I-rule.* There are then three possible cases.

$A$ is $A_1 \& A_2$. Then $\Pi$ has the form $((\Sigma_1/A_1), (\Sigma_2/A_2)/A_1 \& A_2)$. Let $\Delta_j^i$ $(i, j = 1, 2)$ be the formulas in $\Gamma_j$ that the end-formula of $(\Sigma_i/A_i)$ depends on. By the induction assumption, there is an interpolation formula $F_i$ to $\langle \Delta_1^i, \langle \Delta_2^i, A_i \rangle \rangle$, and $F_1 \& F_2$ is then an interpolation formula as desired.

$A$ is $A_1 \supset A_2$. Then $\Pi$ has the form $((\Sigma/A_2)/A_1 \supset A_2)$, where the end-formula of $(\Sigma/A_2)$ depends on $\Gamma$ or on $\Gamma \cup \{A_1\}$. In the first case the interpolation formula to $\langle \Gamma_1, \langle \Gamma_2, A_2 \rangle \rangle$ is of the desired kind. In the second case, there is by the induction assumption an interpolation formula to $\langle \Gamma_1, \langle \Gamma_2 \cup \{A_1\}, A_2 \rangle \rangle$ and this is a desired interpolation formula.

$A$ is $\forall x B$. Then $\Pi$ has the form $((\Sigma/B_a^x)/\forall x B)$. Let $F$ be an interpolation formula to $\langle \Gamma_1, \langle \Gamma_2, B_a^x \rangle \rangle$. Then $\forall x F_x^a$, where $x$ does not occur in $F$, is an interpolation formula as desired. We note that $\Gamma_1 \vdash F$ and that by restriction on the $\forall$I-rule, the parameter $a$ does not occur in any formula of $\Gamma$. Hence $\Gamma_1 \vdash \forall x F_x^a$.

*Case II. The end-formula of $\Pi$ is the consequence of the $\wedge_c$-rule.* Then $\Pi$ has the form $((\Sigma/\wedge)/A)$, where $(\Sigma/\wedge)$ is a deduction of $\wedge$ from $\Gamma$ or from $\Gamma \cup \{\sim A\}$. An interpolation formula of the desired kind is obtained, in the first case, by taking the formula to $\langle \Gamma_1, \langle \Gamma_2, \wedge \rangle \rangle$, and in the second case, by taking the one to $\langle \Gamma_1, \langle \Gamma_2 \cup \{\sim A\}, \wedge \rangle \rangle$.

*Case III. The end-formula of $\Pi$ is the consequence of an E-rule.* Then we consider a main branch $\beta$ in $\Pi$ (i.e. one of order o). Let $C$ be the first formula occurring in $\beta$. By Theorem 3, $C \in \Gamma$. There are again three possible cases.

$C$ is $C_1 \& C_2$. Then $\Pi$ has the form

$$\frac{C_1 \& C_2}{\underset{\Pi^*}{(C_i)}}$$

where $i$ is 1 or 2, and $\Pi^*$ is a deduction of $A$ from $\Gamma^* \cup \{C_i\}$, where $\Gamma^*$ is like $\Gamma$ except for possibly not containing $C$. Let $\Gamma_j^*(j = 1, 2)$ be like

$\Gamma_j$ with the following two exceptions for the case when $C \in \Gamma_j$: (i) $C_i$ is to belong to $\Gamma_j^*$ and (ii) if $C$ does not belong to $\Gamma^*$, then $C$ is not to belong to $\Gamma_j^*$. By the induction assumption, there is an interpolation formula to $\langle\langle \Gamma_1^*, \langle \Gamma_2^*, A \rangle\rangle$ and this is an interpolation formula as desired.

$C$ *is* $C_1 \supset C_2$. Then $\Pi$ has the form

$$\frac{\Sigma_1}{C_1} \qquad C_1 \supset C_2$$
$$\frac{\phantom{}}{(C_2)}$$
$$\Pi^*$$

where $(\Sigma_1/C_1)$ is a deduction of $C_1$ depending on some set $\Delta^1$ such that $\Delta^1 \subset \Gamma$ (cf. Theorem 3) and $\Pi^*$ is a deduction of $A$ depending on $\Delta^2 \cup \{C_2\}$ for some set $\Delta^2 \subset \Gamma$. Let $\Delta_j^i$ $(i, j = 1, 2)$ be $\Gamma_j \cap \Delta^i$.

*Subcase* (i): $C \in \Gamma_1$. Let $F_1$ be an interpolation formula to $\langle \Delta_2^1, \langle \Delta_1^1, C_1 \rangle\rangle$, and let $F_2$ be an interpolation formula to $\langle \Delta_1^2 \cup \{C_2\}, \langle \Delta_2^2, A \rangle\rangle$. Then $F_1 \supset F_2$ is an interpolation formula as desired, which fact is partly verified as follows:

By definition of $F_1$ and $F_2$, we have $\{F_1\} \cup \Delta_1^1 \vdash C_1$ and $\Delta_1^2 \cup \{C_2\} \vdash F_2$. It follows that $\{F_1\} \cup \Gamma_1 \vdash C_1$ and $\Gamma_1 \cup \{C_2\} \vdash F_2$, and since $C_1 \supset C_2 \in \Gamma_1$, it follows that $\{F_1\} \cup \Gamma_1 \vdash F_2$ and hence $\Gamma_1 \vdash F_1 \supset F_2$.

Further, again by definition of $F_1$ and $F_2$, we have $\Delta_2^1 \vdash F_1$ and $\{F_2\} \cup \Delta_2^2 \vdash A$. It follows that $\Gamma_2 \vdash F_1$ and $\{F_2\} \cup \Gamma_2 \vdash A$ and hence $\{F_1 \supset F_2\} \cup \Gamma_2 \vdash A$.

*Subcase* (ii): $C \in \Gamma_2$. Then there is an interpolation formula $F_1$ to $\langle \Delta_1^1, \langle \Delta_2^1, C_1 \rangle\rangle$, and an interpolation formula $F_2$ to $\langle \Delta_1^2, \langle \Delta_2^2 \cup \{C_2\}, A \rangle\rangle$. The desired interpolation formula is $F_1 \& F_2$.

$C$ *is* $\forall x B$. Then $\Pi$ has the form $(\forall x B/B_t^r/\Pi^*)$. $\Pi^*$ is a deduction of $A$ from $\Gamma^* \cup \{B_t^r\}$, where $\Gamma^*$ is like $\Gamma$ except for possibly not containing $C$ (note that there is no premiss of an application of $\forall$I in $\beta$). We define $\Gamma_j^*$ $(j = 1, 2)$ similarly to the way it was defined in the &-case. Let $F$ be an interpolation formula to $\langle \Gamma_1^*, \langle \Gamma_2^*, A \rangle\rangle$. We get the desired interpolation formula as follows. If $C \in \Gamma_1$, we take a universal closure of $F$ with respect to the parameters (if any) that occur in $F$ but not in any formula of $\Gamma_1$. If $C \in \Gamma_2$, we take an existential closure of $F$ with respect to the parameters (if any) that occur in $F$ but not in any formula of $\Gamma_2 \cup \{A\}$.

# IV. NORMAL DEDUCTIONS IN INTUITIONISTIC LOGIC

## § 1. Theorem on normal deductions

Concerning the question of a normal form for deductions in the systems for intuitionistic and minimal logic, we note that applications of the $\vee$E- and $\exists$E-rules give formula occurrences immediately below each other of the same shape. A sequence of formula occurrences of the same shape obtained in that way will be called a *segment*. In a normal deduction we want to exclude the possibility that a segment begins with a consequence of an I-rule or the $\wedge_1$-rule and ends with a major premiss of an E-rule. More precisely, we make the following definitions.

A *segment* in a deduction $\Pi$ is a sequence $A_1$, $A_2$, ..., $A_n$ of consecutive formula occurrences in a thread in $\Pi$ such that

1) $A_1$ is not the consequence of an application of $\vee$E or $\exists$E;

2) $A_i$, for each $i < n$, is a minor premiss of an application of $\vee$E or $\exists$E; and

3) $A_n$ is not the minor premiss of an application of $\vee$E or $\exists$E.

Note that a single formula occurrence that is not the consequence or minor premiss of an application of $\vee$E or $\exists$E is a segment by this definition (clause 2 being vacuously satisfied), and that all formula occurrences in a segment are of the same shape.

A segment $\sigma_1$ is said to be *above* [*below*] a segment $\sigma_2$ if the formula occurrences in $\sigma_1$ stand above [below] the formula occurrences in $\sigma_2$.

A *maximum segment* is a segment that begins with a consequence of an application of an I-rule or the $\wedge_1$-rule and ends with a major premiss of an E-rule. Note that a maximum formula as defined in Chapter III is a special case of a maximum segment.

By a *normal* deduction, I shall now and henceforth (if nothing else is said) mean a deduction that contains no maximum segment and no redundant applications of $\vee$E or $\exists$E. An application of $\vee$E or $\exists$E in a deduction is said to be redundant if it has a minor premiss at which no assumption is discharged; obviously, such applications are superfluous. (Note that a normal deduction in C' is normal also according to this definition.)

We now have:

THEOREM I. *If $\Gamma \vdash A$ holds in the system for intuitionstic or minimal logic, then there is a normal deduction in this system of $A$ from $\Gamma$.*[1]

*Proof.* By the *degree* of a segment, I mean the degree of the formula that occurs in the segment; and by the *length* of a segment, the number of formula occurrences in the segment. The *induction value* of a deduction $\Pi$ is defined as the pair $\langle d, l \rangle$ such that $d$ is the highest degree of a maximum segment in $\Pi$ or o if there is no such segment, and $l$ is the sum of the lengths of maximum segments in $\Pi$ of degree $d$. The induction value $\langle d', l' \rangle$ is *less* than the induction value $\langle d, l \rangle$ if and only if either (i) $d' < d$ or (ii) $d' = d$ and $l' < l$.

Let $\Pi$ be a deduction in I or M of $A$ from $\Gamma$ that has only pure parameters, Let $v = \langle d, l \rangle$ be its induction value and assume that $d > 0$. We shall show that there is a deduction $\Pi'$ in I or M (respectively) of $A$ from $\Gamma$ that has an induction value less than $v$. This proves the theorem, since the induction value is not changed by making the proper parameters pure and since it is trivial to remove redundant applications of $\vee$ E and $\exists$ E.

To prove the assertion, we choose a maximum segment $\sigma$ of degree $d$ in $\Pi$ such that there is (i) no maximum segment of degree $d$ above $\sigma$ and (ii) no maximum segment of degree $d$ that stands above or contains a formula occurrence side-connected with the last formula occurrence in $\sigma$.[2]

If $\sigma$ is a maximum formula that is the consequence of an I-rule, then let $\Pi'$ be a reduction of $\Pi$ at $\sigma$ as defined in Chapter II. It is easily seen that the induction value of $\Pi'$ is less than $v$. The case when $\sigma$ is a maximum formula and consequence of the $\wedge_1$-rule is treated simi-

---

[1] This theorem can be proved for the system C, too. Theorem I in Ch. III, however, cannot be extended to C.

[2] That there exists a segment of this kind can easily be seen., e.g. as follows. Consider the set of maximum segments of degree $d$ that satisfy clause (i). If $\sigma_1$ is a segment in this set for which clause (ii) does not hold, then there is obviously a segment $\sigma_2$ in this set that makes clause (ii) fail for $\sigma_1$. If clause (ii) also fails for $\sigma_2$, then we can find a third segment $\sigma_3$ in the set that makes clause (ii) fail for $\sigma_2$, etc. In this way, we obtain a sequence $\sigma_1, \sigma_2, \sigma_3, \ldots$ and clearly $\sigma_i \neq \sigma_j$ if $i \neq j$. Thus, we must finally get a segment in the set that also satisfies clause (ii).

larly (but is easier). If the length of $\sigma$ is greater than one, we reduce the length of $\sigma$ as follows. Let $F$ be the formula that occurs in $\sigma$. There are two possible cases:

*The last occurrence of $F$ in $\sigma$ is the consequence of $\vee$E.* Then $\Pi$ has the form shown to the left below, except that $\Sigma_4$ may instead stand to the left of $F$, but that case is quite symmetrical to the one below ($\Sigma_4$ may, of course, also be empty). $\Pi'$ is to be as shown to the right below and is thus obtained by moving the application of $\vee$E one step down, so to say.

$$
\begin{array}{ccc}
\dfrac{\Sigma_1}{B \vee C} \quad \dfrac{\Sigma_2}{F} \quad \dfrac{\Sigma_3}{F} & & \\[4pt]
\dfrac{\phantom{X}}{F} \qquad \Sigma_4 & & \\
(C) & & \\
\Pi_5 & &
\end{array}
\qquad
\begin{array}{c}
\dfrac{\Sigma_1}{B \vee C} \quad \dfrac{\dfrac{\Sigma_2}{F}\ \Sigma_4}{D} \quad \dfrac{\dfrac{\Sigma_3}{F}\ \Sigma_4}{D} \\[4pt]
(C) \\
\Pi_5
\end{array}
$$

We note that the application of the E-rule in $\Pi$ that has the last occurrence of $F$ in $\sigma$ as major premiss is replaced in $\Pi'$ by two applications of the same form, but with the difference that the major premisses of these applications now depend also on the formulas $B$ and $C$ respectively. However, this is inessential since the restrictions concerning the formulas that a formula occurrence depends on do not concern major premisses of E-rules, (but only premisses of $\forall$I and minor premisses of $\exists$E). One can then see that $\Pi'$ is still a deduction of $A$ from $\Gamma$. That the induction value of $\Pi'$ is less than $v$ is obtained by applying clause (ii) in the assumptions about $\sigma$.

*The last occurrence of $F$ in $\sigma$ is the consequence of $\exists$E.* Then $\Pi$ has the form shown to the left below, again with the possible exception that $\Sigma_4$ may stand to the right of $F$ instead. $\Pi'$ is to be as shown to the right below.

$$
\begin{array}{cc}
\dfrac{\Sigma_1}{\exists x B} \quad \dfrac{\Sigma_2}{F} & \\[4pt]
\dfrac{\phantom{X}}{F} \quad \Sigma_4 & \\
(D) & \\
\Pi_5 &
\end{array}
\qquad
\begin{array}{c}
\dfrac{\Sigma_1}{\exists x B} \quad \dfrac{\dfrac{\Sigma_2}{F}\ \Sigma_4}{D} \\[4pt]
(D) \\
\Pi_5
\end{array}
$$

Note that the proper parameter of the application of $\exists$E in question does not occur in $\Sigma_4$ (clause 2 in the lemma on parameters). That $\Pi'$ is still a deduction of $A$ from $\Gamma$ then follows as above. The induction value of $\Pi'$ is clearly less than $v$.

## § 2. The form of normal deductions

What was said in Chapter III about branches in a normal deduction
now holds if we take the segments, instead of the formula occurrences,
as units and replace the branches with sequences of formula occurrences which are like branches except that the formula occurrence immediately succeeding a major premiss of an application $\alpha$ of $\vee$E or
$\exists$E is an assumption discharged by $\alpha$ instead of the consequence of $\alpha$.
Such a sequence is called a *path* if it does not begin with an assumption
discharged by an application of $\vee$E or $\exists$E. More precisely, we define:

A sequence $A_1, A_2, ..., A_n$ is a *path* in the deduction $\Pi$ if and only if

1) $A_1$ is a top-formula in $\Pi$ that is not discharged by an application
of $\vee$E or $\exists$E; and

2) $A_i$, for each $i < n$, is not the minor premiss of an application of
$\supset$E, and either (i) $A_i$ is not the major premiss of $\vee$E or $\exists$E, and
$A_{i+1}$ is the formula occurrence immediately below $A_i$, or (ii) $A_i$ is the
major premiss of an application $\alpha$ of $\vee$E or $\exists$E, and $A_{i+1}$ is an assumption discharged in $\Pi$ by $\alpha$; and

3) $A_n$ is either a minor premiss of $\supset$E, the end-formula of $\Pi$, or a
major premiss of an application $\alpha$ of $\vee$E or $\exists$E such that $\alpha$ does not
discharge any assumptions.

As an *example*, we have three paths in the deduction of p. 19, namely
  (i) the sequence of the formula occurrences marked (1), (3), left
       occurrence of (4), (7), (8), (9), (10), and
  (ii) the sequence of those marked (1), (3), right occurrence of (4),
       and
  (iii) the thread that starts with the formula occurrence marked (2).

In normal deductions, the last formula occurrence in a path is
always a minor premiss of $\supset$E or the end-formula of the deduction,
since the possibility of applications of $\vee$E and $\exists$E that do not discharge any assumption is then excluded.

Every path $\pi$ can obviously be uniquely divided into consecutive
segments (usually consisting of just one formula occurrence). In other
words, $\pi$ is the concatenation of a sequence of segments and can be
written

$$A_{1,1}, A_{1,2}, ..., A_{1,n_1}, A_{2,1}, A_{2,2}, ..., A_{2,n_2}, ..., A_{k,1}, A_{k,2}, ..., A_{k,n_k},$$

where $A_{i,1}, A_{i,2}, ..., A_{i,n_i}$ for each $i \leqslant k$ is a segment $\sigma_i$ in $\Pi$. The
sequence $\sigma_1, \sigma_2, ..., \sigma_k$ will be called *the sequence of segments in $\pi$*.

It will be convenient to say that a segment $\sigma$ is a *top-segment* [*end-segment*] or is a *consequence* [(*major* or *minor*) *premiss*] of an application $\alpha$ of an inference rule when the first [last] formula occurrence in $\sigma$ is a top-formula [the end-formula] or a consequence [(major or minor) premiss] of $\alpha$ respectively. However, this terminology will be used only in phrases such as "the segment $\sigma$ is a consequence of $\alpha$" or "the segment $\sigma$ that is premiss of $\alpha$"; when not explicitly speaking about segments, we still mean by the premiss or the consequence of an application of a rule the formula occurrence in question.

THEOREM 2. *Let* $\Pi$ *be a normal deduction in* I *or* M, *let* $\pi$ *be a path in* $\Pi$, *and let* $\sigma_1, \sigma_2, ..., \sigma_n$ *be the sequence of segments in* $\pi$. *Then there is a segment* $\sigma_i$, *called the* minimum segment *in* $\pi$, *which separates two (possibly empty) parts of* $\pi$, *called the* E-part *and* I-part *of* $\pi$, *with the properties:*

1) *For each* $\sigma_j$ *in the E-part (i.e.* $j < i$) *it holds that* $\sigma_j$ *is a major premiss of an E-rule and that the formula occurring in* $\sigma_j$ *is a subformula of the one occurring in* $\sigma_{j+1}$.

2) $\sigma_i$, *provided that* $i \neq n$, *is a premiss of an I-rule or of the* $\bigwedge_1$-*rule.*[1]

3) *For each* $\sigma_j$ *in the I-part, except the last one, (i.e.,* $i < j < n$) *it holds that* $\sigma_j$ *is a premiss of an I-rule and that the formula occurring in* $\sigma_j$ *is a subformula of the one occurring in* $\sigma_{j+1}$.

We assign an *order* to the paths in the same way as to the branches. A path of order o is said to be a *main path*. We then prove as in Chapter II:

COROLLARY 1. (Subformula principle.) *Every formula occurring in a normal deduction in* I *or* M *of* $A$ *from* $\Gamma$ *is a subformula of* $A$ *or of some formula of* $\Gamma$.

This corollary can be strengthened as in Chapter III. To clause 2) in the definition of positive and negative subformula, we add:

$B$ and $C$ are positive [negative] subformulas of $A$ when $B \vee C$ is one, and so is $B_t^x$ when $\exists x B$ is.

We then have:

---

[1] In the case of minimal logic, $\sigma_i$ can of course be a premiss of an I-rule only; the formula occurring in $\sigma_i$ is then always a subformula of the one occurring in $\sigma_{i+1}$ (provided that $i \neq n$).

COROLLARY 2. *Corollary 2 in Chapter II holds also for* I *and* M *but now with the following simplifications:*

  (i) *The second part of clause 1) concerning* $\wedge_c$ *is left out.*

  (ii) *In the case of minimal logic, the phrase "if different from* $\wedge$*" (in clause 2) is left out.*

As a lemma for later theorems, we record also the following properties concerning branches in normal deductions in I and M (it can be obtained as a corollary from Theorem 2 or proved directly in the same way as that theorem):

CORALLARY 3. *Let* $\Pi$ *be a normal deduction in* I *or* M *and let* $A_1$, $A_2$, ..., $A_n$ *be a branch in* $\Pi$ *that contains no minor premisses of* $\vee$E *or* $\exists$E. *Then there is a formula occurrence* $A_i$ *such that:*

  1) $A_j$, *for each* $j < i$, *is a major premiss of an E-rule;*

  2) $A_i$, *provided* $i \neq n$, *is a premiss of an I-rule or the* $\wedge_1$*-rule;*

  3) $A_j$, *for each* $i < j < n$, *is a premiss of an I-rule and is a subformula of* $A_{j+1}$.

Finally, we note the following corollary, whose proof is immediate from Corollary 1 by inspection of the inference rules:

COROLLARY 4. (Separation theorem.)[1] *If* $\Pi$ *is a normal deduction in* I *or* M *of A from* $\Gamma$, *then the only inference rules that are applied in* $\Pi$ *are the inference rules in* I *or* M *for the logical constants that occur in A or in some formula of* $\Gamma$.

# § 3. Some further corollaries

We note some further results for the systems for intuitionistic and minimal logic that can be obtained as corollaries from the theorem on normal deductions. Some of them are already well known from other or similar proofs but are obtained here by relatively easy and uniform applications of Theorems 1 and 2. What is said in this section is to be

---

[1] A similar theorem for the calculus of sequents is an immediate corollary of Gentzen's Hauptsatz (cf. App. A). For a system of intuitionistic logic of axiomatic type, a similar result was stated by Wajsberg [1] and proved by Curry [1]. An algebraic proof of the separation theorem for the sentential part of intuitionistic logic has recently been given by Horn [1].

understood as referring to the system for intuitionistic logic or, alternatively, to the system for minimal logic.[1]

COROLLARY 5. *The interpolation theorem (Corollary 5 in Chapter III) holds also for intuitionistic and minimal logic.*[2]

The proof is obtained from the proof of Corollary 5 in Chapter III by adding cases for ∨ and ∃. (They become more or less dual to the ones for & and ∀.) In case II, we consider the $\wedge_i$-rule instead of the $\wedge_c$-rule. In case III, we consider a thread $\tau$ that contains no minor premiss (instead of the main branch $\beta$) and apply Corallary 3 (instead of Theorem 3).

COROLLARY 6. *If $\Gamma \vdash A \vee B$, then either $\Gamma \vdash A$ or $\Gamma \vdash B$, provided that no formula of $\Gamma$ has a strictly positive subformula that contains ∨ as principal sign.*[3]

*Proof.* Let $\Pi$ be a normal deduction of $A \vee B$ from $\Gamma$. We first show that there is exactly one end-segment $\sigma$ in $\Pi$. If there were two end-segments, they must contain a minor premiss of ∨E. Consider the corresponding major premiss $C \vee D$ and let $\pi$ be a path to which $C \vee D$ belongs. The first formula occurrence $F$ in $\pi$ cannot be discharged in $\Pi$ (since $C \vee D$ belongs to the E-part of $\pi$ and there is no premiss of $\supset$ I below $C \vee D$) and hence $F \in \Gamma$. But by Corollary 2, $C \vee D$ is then a strictly positive subformula of $F$ contrary to the assumptions about $\Gamma$.

$\sigma$ must further be consequence of an application $\alpha$ of an I-rule (i.e. ∨I) or of the $\wedge$-rule; otherwise, it is a minimum segment of the

---

[1] The proofs are usually given with the case of intuitionistic logic in mind and could be simplified in case of minimal logic.

[2] First proved by Schütte [3] (cf. note 1 on p. 46).

[3] For the case when $\Gamma$ is empty, the theorem was first stated (without proof and for the propositional calculus only) by Gödel [2] and proved by Gentzen [3] (p. 407) as a corollary of his Hauptsatz. Other proofs are found in e.g., McKinsey-Tarski [1] and Harrop [1].

With the present restriction of $\Gamma$, the theorem was proved by Harrop [2] for the intuitionistic propositional calculus and for the intuitionistic elementary number theory; in the latter case provided that all the formulas involved are closed (or, in our terminology, that they do not contain individual parameters). Harrop's proof is not constructive in the case of number theory. Kleene [2] extends this result to intuitionistic predicate logic and considers also a weaker but not effective restriction on $\Gamma$.

paths to which it belongs, and we then obtain as above that the first formula that occurs in any of these paths belongs to $\Gamma$ and has $A \vee B$ as strictly positive subformula. It follows that the premiss of $\alpha$ is $A$, $B$, or $\wedge$. For simplicity, we may assume that it is $A$ or $B$, since instead of inferring $A \vee B$ directly from $\wedge$, we can first infer $A$, e.g. It is then clear how to obtain a deduction of $A$ or $B$ respectively from $\Gamma$: We just refrain from inferring $A \vee B$ by the application $\alpha$ and substitute $A$ or $B$, respectively, for all occurrences of $A \vee B$ in $\sigma$. In more detail for the case when $\sigma$ contains more than one element: Let $(\Sigma/A)$ be the subdeduction of $\Pi$ that has the major premiss of $\alpha$ as end-formula. Then the given deduction and the new deduction have respectively the forms

$$
\begin{array}{cc}
 & \dfrac{\Sigma}{A} \\
\dfrac{\Sigma_1}{\exists x_1 C_1} & A \vee B \\
\hline
\dfrac{\Sigma_2}{\exists x_2 C_2} & A \vee B \\
\vdots \\
\dfrac{\Sigma_n}{\exists x_n C_n} & A \vee B \\
\hline
A \vee B
\end{array}
\qquad
\begin{array}{cc}
\dfrac{\Sigma_1}{\exists x_1 C_1} & \dfrac{\Sigma}{A} \\
\hline
\dfrac{\Sigma_2}{\exists x_2 C_2} & A \\
\vdots \\
\dfrac{\Sigma_n}{\exists x_n C_n} & A \\
\hline
A
\end{array}
$$

COROLLARY 7.[1] *Let* $t_1, t_2, ..., t_n$ $(n \geqslant 0)$ *be all the terms that occur in* $\exists x A$ *or in some formula of* $\Gamma$ *and let there be no formula of* $\Gamma$ *that has a strictly positive subformula containing* $\exists$ *as principal sign.*

*We then have:*

(i) *For* $n > 0$: *If* $\Gamma \vdash \exists x A$, *then* $\vdash A_{t_1}^x \vee A_{t_2}^x \vee ... \vee A_{t_n}^x$.

(ii) *For* $n > 0$ *and provided that no formula of* $\Gamma$ *has a strictly positive subformula that contains* $\vee$ *as principal sign: If* $\Gamma \vdash \exists x A$, *then* $\Gamma \vdash A_{t_i}^x$ *for some* $i \leqslant n$.

(iii) *For* $n = 0$: *If* $\Gamma \vdash \exists x A$, *then* $\Gamma \vdash \forall x A$.

*Proof.* Let $\Pi$ be a normal deduction of $\exists x A$ from $\Gamma$. By the lemma on parameters, we can assume that the proper parameters are pure and different from the ~~terms~~ *parameter* $t_1, t_2, ..., t_n$ that occur in $\exists x A$ or in some formula of $\Gamma$. Let $\sigma$ be any end-segment in $\Pi$. We then have that

---

[1] For the case when $\Gamma$ is empty, part (ii) of the corollary can be obtained from Gentzen [3].

A result similar to part (ii) of the corollary is obtained by Harrop [2] for intuitionistic elementary number theory (the term $t_i$ is then a numeral). Kleene [2] and [3] prove part (ii) of the corollary also for predicate logic and consider also a weaker but not effective restriction on $\Gamma$.

(1)                    $\sigma$ contains no minor premiss of $\exists$E

and that $\sigma$ is the consequence of an application $\alpha$ of an I-rule (i.e. $\exists$I) or the $\wedge_1$-rule; otherwise, as in the proof of Corollary 6, we obtain that a formula of $\Gamma$ has a strictly positve subformula that contains $\exists$ as principal sign. Hence, the major premiss of $\alpha$ has the form $A_u^x$ or $\wedge$. For simplicity, we may assume that it has the form $A_u^x$, since instead of inferring $\exists xA$ directly from $\wedge$ we can first infer $A_u^x$. If $u$ is not one of $t_1, t_2, ..., t_n$, we now replace all occurrences of $u$ in $\Pi$ by one cf the terms $t_i$ (e.g. $t_1$) in case $n > 0$ and by a parameter $a$ not occurring in $\Pi$ in case $n = 0$. We repeat this process for every end-segment of $\Pi$ so that every one comes to stand immediately below a formula occurrence of one of the two forms $A_{t_i}^x$ (for some $i \leqslant n$) and $A_a^x$. Then, we still have a deduction $\Pi'$ of $\exists xA$ from $\Gamma$ (note that neither of $u$, $a$, and $t_i$ $(i \leqslant n)$ is a proper parameter in $\Pi$).

We first prove part (ii) of the corollary. In this case, we know from the proof of Corollary 6 that there is exactly one end-segment $\sigma$ in $\Pi'$. Hence, by (1), $\sigma$ consists of just the end-formula $\exists xA$ of $\Pi'$, and leaving out this end-formula, we have a deduction of $A_{t_i}^x$ from $\Gamma$ for some $i \leqslant n$.

In case (iii), every end-segment in $\Pi'$ stands immediately below a major premiss $A_a^x$ of an application of $\exists$I. As in the proof of Corollary 6 —except that we may now have several end-segments and a series of applications of $\vee$E instead of $\exists$E—we get a deduction $\Pi^*$ of $A_a^x$ from $\Gamma$ by leaving out these applications of $\exists$I and substituting $A_a^x$ for all occurrences of $\exists xA$ in the end-segments of $\Pi'$. As $a$ does not occur in any assumptions on which the end-formula $A_a^x$ of $\Pi^*$ depends, we can then apply $\forall$I and obtain the desired deduction: $(\Pi^*/\forall xA)$.

In case (i), finally, every end-segment in $\Pi'$ stands immediately below a premiss $A_{t_i}^x$ of an application of $\exists$I. We leave out these applications of $\exists$I, insert instead a series of applications of $\vee$I ending with $A_{t_1}^x \vee A_{t_2}^x \vee ... \vee A_{t_n}^x$, and substitute this disjunction for all occurrences of $\exists xA$ in the end-segments of $\Pi'$. Because of (1), this does not conflict with the restrictions on $\exists$E, and we thus have the desired deduction.

COROLLARY 8. *Let $C$ be a formula which does not contain any occurrence of $\supset$, and let $\Gamma$ be $\{A_1 \supset B_1, A_2 \supset B_2, ..., A_n \supset B_n\}$. If $\Gamma \vdash C$, then $\Gamma \vdash A_i$ for some $i \leqslant n$.*[1]

---

[1] Cf. Wajsberg [1].

*Proof.* Let $\Pi$ be a normal deduction of $C$ from $\Gamma$ and let $\tau$ be a thread (main branch) in $\Pi$ that contains no minor premiss. As $C$ contains no occurrence of $\supset$, it follows from Corollary 3 that no assumption is discharged in $\Pi$ at a formula occurrence in $\tau$. Hence, the first formula occurrence in $\tau$ has the form $A_i \supset B_i$ (for some $i \leqslant n$), and $\Pi$ has the form

$$\frac{\dfrac{\Sigma}{A_i} \quad A_i \supset B_i}{\begin{array}{c}[B_i]\\ \Pi'\end{array}}$$

where $A_i$ depends on formulas of $\Gamma$ only. Hence, $(\Sigma/A_i)$ is a deduction of $A_i$ from $\Gamma$.

For the next corollary, I shall say that a binary sentential connective [quantifier] $\gamma$ is *(strongly) definable* in the system S if for every formula $A$ and $B$ there is a formula $C$ not containing $\gamma$ such that $\vdash_s (A\gamma B) \equiv C$ [$\vdash_s \gamma x A \equiv C$].[1] I shall say that $\wedge$ is (strongly) definable in S if there is a formula $A$ not containing $\wedge$ such that $\vdash_s \wedge \equiv A$.

There is also a weaker sense in which a logical constant may be definable in a system S, namely when we can uniformly transform every formula containing $\gamma$ to a formula not containing $\gamma$ without changing provability in S. More precisely, I shall say that a binary sentential connective [quantifier] $\gamma$ is *weakly definable* in S if there is a transformation that transforms any pseudo-formula $A$ to a pseudo-formula $A^*$ not containing $\gamma$ in such a way that:

(i) $A^*$ can be obtained from $A$ as the result of successively replacing parts of the form $(B \gamma C)$ $[\gamma x B]$ by $(B \gamma C)^*$ $[(\gamma x B)^*]$;[2] and

(ii) $\qquad\qquad \vdash_s A^*$ if and only if $\vdash_s A$.

Similarly, I shall say that $\wedge$ is weakly definable in S if there is a formula $B$ not containing $\wedge$ such that if $A^*$ is obtained from $A$ by replacing every occurrence of $\wedge$ with $B$, then $\vdash_s A^*$ if and only if $\vdash_s A$.

---

[1] Actual definitions of course also satisfy the stronger requirement that $C$ is determined uniformly by $A$ and $B$ [by $A$] (e.g. by a schema). However, I shall here be concerned with showing non-definability, and that will be possible without imposing such a requirement.

[2] Again, most transformations of this kind that have actually been used are uniform in a stronger sense than required by (i) (cf. note 1 above).

Of course, strong definability implies weak definability. Conversely, it can rather easily be seen that weak definability of one of the constants &, v, and E in the system for intuitionistic or minimal logic implies strong definability. However, $\wedge$ is weakly, but not strongly, definable in the system for minimal logic. Let $P$ be a 0-place predicate parameter and let $A^*$ be the result of replacing every occurrence of $\wedge$ in $A$ with $P$. It can then easily be seen that $\vdash_M A^*$ if and only if $\vdash_M A$.

I shall say that a logical constant is (*weakly*) *independent* [*strongly independent*] in S if $\gamma$ is not strongly [weakly] definable in S.

CORALLARY 9. *All the logical constants are strongly independent in the system of intuitionistic logic. In the system for minimal logic, $\wedge$ is weakly independent, and all the other logical constants are strongly independent.*[1]

For the proofs below, we assume—contrary to what is to be proved—that there is a transformation (*) as required in the definition of weak definability.

*Independence of* &. Let $P$ and $Q$ be different 0-place predicate parameters, and let $(P \& Q)^* = A$. Since $\vdash (P \& Q) \supset P$ and $\vdash (P \& Q) \supset Q$, we have $\vdash A \supset P$ and $\vdash A \supset Q$, and hence $A \vdash P$ and $A \vdash Q$. We prove the following general lemma:

*If $A$ lacks occurrences of & and $P \neq Q$, then $A \vdash P$ and $A \vdash Q$ cannot both hold unless $A \vdash \wedge$.*

A contradiction is then obtained, e.g. as follows: $\vdash P \supset (Q \supset (P \& Q))$, hence $\vdash P \supset (Q \supset A)$ and, by the lemma, $\{P, Q\} \vdash \wedge$. The latter can easily be seen to be false by application of Theorems 1 and 2.

To prove the lemma, let $\Pi_1$ [$\Pi_2$] be a normal deduction of $P$ [$Q$] from $A$, and let $\pi_1$ be a main path in $\Pi_1$. Thus, $\pi_1$ begins with an occurrence of $A$ and ends with an occurrence of $P$. As $P$ is atomic, the minimum segment of $\pi_1$ consists either of occurrences of $P$, in which case the I-part of $\pi_1$ is empty, or of occurrences of $\wedge$, in which case the I-part of $\pi_1$ contains occurrences of $P$ only. Now consider any two paths in normal deductions that begin with assumptions of the same form lacking occurrences of &, and consider the formulas that

[1] That none of the constants &, $\vee$, $\supset$, and $\sim$ can be (strongly) defined in terms of the other three is known from Wajsberg [1] and McKinsey [1]. As the present corollary shows, they are still independent (also in a stronger sense) when first order quantifiers are added. However, in second order systems, we shall see that &, $\vee$, $\exists$, and $\wedge$ are (strongly) definable in terms of $\supset$ and $\forall$ (Ch. V).

occur in the successive segments in these paths. It is then easily verified that the first place at which these formulas can differ—disregarding differences with respect to individual parameters—is either at segments that immediately succeed segments that are major premisses of $\vee E$ or at segments one of which belongs to the I-part. Thus, if $\pi_2$ is a main path in $\Pi_2$ such that the choices with respect to successors to segments that are major premisses of $\vee E$ are made in the same way as in $\pi_1$, then the formula occurring in the minimum segment of $\pi_2$ must be identical with the formula occurring in the minimum segment of $\pi_1$. As $P \neq Q$, it follows that every main path in $\Pi_1 [\Pi_2]$ has a minimum segment that consists of occurrences of $\wedge$ and that is a premiss of an application of the $\wedge_1$-rule, having $P [Q]$ as consequence. Leaving out these applications of the $\wedge_1$-rule in $\Pi_1$ (cf. the proof of Corollary 6 or 7) and substituting $\wedge$ for $P$ in all end-segments of $\Pi_1$, we obtain a deduction of $\wedge$ from $A$.

*Independence of* $\vee$. Let $P$ and $Q$ be different o-place parameters, and let $(P \vee Q)^* = A$. Let $R$ be a o-place parameter that does not occur in $A$. Since

$$\vdash (P \vee Q) \supset ((P \supset R) \& (Q \supset R) \supset R)$$

it follows that $A \vdash (P \supset R) \& (Q \supset R) \supset R$ and hence that

$$(1) \qquad\qquad A \vdash P \vee Q$$

(substitute $(P \vee Q)$ for $R$). Corollary 6 applied to (1) gives that

$$(2) \qquad\qquad \text{either } A \vdash P \text{ or } A \vdash Q.$$

But, on the other hand, $\vdash P \supset (P \vee Q)$ and $\vdash Q \supset (P \vee Q)$, and hence $P \vdash A$ and $Q \vdash A$. Combined with (2), this is absurd, since, by easy application of Theorems 1 and 2, it can be seen that neither $P \vdash Q$ nor $Q \vdash P$.

*Independence of* $\supset$. Immediate, since a formula not containing $\supset$ cannot be provable. (That this is so can be seen from Theorem 2, because if the assumption with which a main path in a normal deduction begins is discharged (as it must be when the deduction is a proof), then the last formula of the path contains an occurrence of $\supset$.)

*Independence of* $\forall$. Let $P$ be a 1-place predicate parameter, and let $(\forall x P x)^* = A$. Let $a$ be a parameter that does not occur in $A$. Since

$\vdash \forall x Px \supset Pa$, also $\vdash A \supset Pa$, and hence $A \vdash \forall x Px$. The following lemma holds in general:

*If $A$ does not contain $\forall$, then $A \vdash \forall x Px$ implies $A \vdash \curlywedge$.*

The independence of $\forall$ then easily follows. (For instance, since $\forall x Px \supset Q$ is not provable, $A \supset Q$ is not provable, contradicting the result that $A \vdash \curlywedge$.)

To prove the lemma, let $\Pi$ be a normal deduction of $\forall x Px$ from $A$ that has only pure parameters, and let $\pi_1$ be a main path in $\Pi$. The minimum segment $\sigma$ of $\pi_1$ consists either of occurrences of $\curlywedge$ or of occurrences of $Pa$ for some $a$. In the latter case, $\sigma$ is the premiss of an application $\alpha$ of $\forall I$. However, at which point in $\pi$ can $a$ enter? It holds in general that an individual parameter can enter into a path only (i) at the first element of the path, (ii) at the consequence of an I-rule, the $\curlywedge_1$-rule or the $\forall E$-rule, or (iii) at an assumption discharged by $\exists E$. Case (iii) is excluded. Since $\Pi$ has only pure parameters, the assumption in question would have to be discharged at some formula occurrence below $\sigma$, and the premiss of $\alpha$ would then depend on an assumption containing $a$, violating the requirements in the deduction rule for $\forall I$. That case (ii) is excluded follows from Theorem 2 and the fact that $A$ lacks occurrences of $\forall$: The minimum segment $\sigma$ cannot be preceded by a consequence of an I-rule or the $\curlywedge_1$-rule, and if $\sigma$ is preceded by a premiss of the $\forall E$-rule, then $\forall$ would occur in the first formula occurrence in $\pi$ i.e., in $A$. But also case (i) is impossible, because the premiss of $\alpha$ would then again have to depend on a formula containing $a$, violating the requirements in the deduction rule for $\forall I$. It follows that every main path has a minimum segment consisting of occurrences of $\curlywedge$, and the lemma then follows as in the &-case above.

*Independence of $\exists$.* Let $P$ be a 1-place parameter, and let $(\exists x Px)^* = A$. Let $Q$ be a 0-place parameter not occurring in $A$. Since

$$\vdash \exists x Px \supset (\forall x(Px \supset Q) \supset Q)$$

it follows that $A \vdash \forall x(Px \supset Q) \supset Q$, and hence that

$$(1) \qquad\qquad A \vdash \exists x Px$$

(substitute $\exists x Px$ for $Q$). Corollary 7 applied to (1) gives that

$$(2) \qquad\qquad A \vdash Pt_1 \vee Pt_2 \vee \dots \vee Pt_n$$

for certain terms $t_i$. Let $a$ be different from all the $t_i$s $(i \leqslant n)$. Since $\vdash Pa \supset \exists x Px$, it follows from (2) that

$$Pa \vdash Pt_1 \vee Pt_2 \vee \ldots \vee Pt_n,$$

which is absurd (apply Corollary 6 and then Theorem 2).

*Weak independence of* $\wedge$. $A \vdash \wedge$ cannot hold if $A$ lacks occurrences of $\wedge$. Because, by Theorem 2, normal deductions contain occurrences of $\wedge$ in minimum segments only, and, by Corollary 2, the formula occurring in the minimum segment in a normal deduction of $\wedge$ from $A$ is an assumption-part of $\langle \{A\}, \wedge \rangle$ and is hence a subformula of $A$.

*Strong independence of* $\wedge$ *in intuitionistic logic.* Suppose that $\wedge$ could be replaced by $A$ without changing provability, and let $P$ be a o-place parameter not occurring in $A$. Since $\vdash_1 \wedge \supset P$, it follows that $A \vdash_1 P$, which is impossible just as above.

# V. SECOND ORDER LOGIC

## § 1. Natural deduction for second order logic

Systems of natural deduction for second order logic are obtained from the systems of Chapter I by a few additions. I shall consider systems both for simple and ramified predicate logic of second order and shall in both cases consider two alternative versions of such systems.

LANGUAGES FOR (SIMPLE) 2ND ORDER LOGIC. The languages for the systems for (simple) second order logic are obtained from the languages of Chapter I by adding predicate variables and allowing quantification also over these variables. The languages for the second version of these systems are also to contain an abstraction operator to form names of $n$-ary relations. In more detail, the additions are as follows.

*Version 1.* To the non-descriptive symbols, we add a denumerable set of *n-place predicate variables*, for each $n \geqslant 0$. As syntactic notation, we use $X^n$ and $Y^n$ (sometimes omitting the superscript).

To the previous definition of the formulas, we add a fourth clause:

4) If $A$ is a formula in $L$, then so are $\forall X^n A^*$ and $\exists X^n A^*$, where $A^*$ is $A$ or is obtained from $A$ by replacing occurrences of an $n$-place predicate parameter by $X^n$.

The definition of a bound and free variable occurrence in a pseudo-formula and the notation for substitution are extended to predicate variables, parameters, and constants in an analogous fashion.

*Version 2.* Besides the addition of the $n$-place predicate variables, we add an *operator $\lambda$ for relation abstraction* as a new logical constant.

By a simulataneous induction, we single out two categories of symbol strings, namely the *formulas* and the *n-place predicate terms*:

0) $\lambda$ is a formula in $L$.

1) If $T$ is an $n$-place predicate term in $L$ and $t_1, t_2, ..., t_n$ are individual terms in $L$, then $Tt_1 t_2 ... t_n$ is a formula in $L$ $(n \geqslant 0)$.—To facilitate

reading when $T$ is formed by the use of $\lambda$ (cf. clause 6 below), I shall sometimes enclose the term in braces writing $\{T\}t_1\, t_2\, ... \, t_n$.

2) and 3) to read as in the definition of formula in Chapter I.

4) to read as in version 1 above.

5) An $n$-place predicate parameter or constant in $L$ is an $n$-place predicate term in $L$.

6) If $A$ is a formula in $L$, then $\lambda x_1\, x_2\, ... \, x_n A^*$ is an $n$-place predicate term in $L$ $(n \geqslant 0)$, where $A^*$ is $A$ or is obtained from $A$ by replacing some occurrences of individual parameters with the variables $x_1$, $x_2$, ..., $x_n$ which are all to be different. When $n = 0$, we may leave out $\lambda$.

The atomic formulas are as before (i.e., they are formed by clause 0 or 1 using a predicate parameter or constant for $T$).

$\lambda$ is the *principal sign* of a predicate term of the form $\lambda x_1\, x_2\, ... \, x_n A$.

The letters $T^n$ and $U^n$ will be used to represent $n$-place predicate terms (sometimes without superscript).

By a pseudo-formula, we mean as before a symbol string that is obtained from a formula by replacing parameters with variables of the same kind. An occurrence of a predicate variable in a pseudo-formula is defined as free or bound as in Chapter I. An occurrence of an individual variable $x$ in a pseudo-formula $A$ is *bound* or *free* in $A$ according as to whether the occurrence belongs or does not belong to the scope of an occurrence in $A$ of a quantifier or $\lambda$ that is immediately followed by the variable $x$ or by a string of variables among which are $x$.

The notation for substitution of individual variables and terms is then defined as before and is extended to predicate variables and terms. Thus, e.g. $A_{U^n}^{T^n}$ is the result of replacing every occurrence of $T^n$ in $A$ by $U^n$, and $A_{T^n}^{X^n}$ is the result of replacing every free occurrence of $X^n$ in $A$ with $T^n$.

LANGUAGES FOR RAMIFIED 2ND ORDER LOGIC. The languages for the systems for ramified second order logic are like the languages for simple second order logic with the sole exception that the predicate variables, parameters and constants are to be divided into categories not only with respect to the number of argument-places but also with respect to *levels*. The levels are numbered 1, 2, ... Thus, for each $n \geqslant 0$ and for each $m > 0$, there is to be a denumerable set of *$n$-place predicate variables of level $m$*. We distinguish two versions of languages for ramified second order logic in the same way as for simple second order logic.

INFERENCE RULES FOR (SIMPLE) 2ND ORDER LOGIC. We state introduction and elimination rules also for quantifiers with predicate variables. To distinguish them from the corresponding rules for quantifiers with individual variables, we call the new rules $\forall^2 I$, $\forall^2 E$, $\exists^2 I$, and $\exists^2 E$. The $\forall^2 E$- and $\exists^2 I$-rules become simpler in the second version of the respective systems, for which we also state an introduction and an elimination rule for $\lambda$. The terminology introduced in Chapter I in connection with inference rules is extended to the new rules in the natural way.

*Version 1.* The new rules are indicated by the figures:

$$\forall^2 I) \quad \frac{A}{\forall X^n A_{X^n}^{P^n}} \qquad\qquad \forall^2 E) \quad \frac{\forall X^n A}{A^*}$$

$$\exists^2 I) \quad \frac{A^*}{\exists X^n A} \qquad\qquad \exists^2 E) \quad \frac{\exists X^n A \quad \overset{(A_P^{X^n})}{B}}{B},$$

where $A^*$ is to fulfill the condition: There is a formula $C$ and parameters $a_1, a_2, ..., a_n$ all different ($n \geqslant 0$) such that (i) $A^*$ is obtained from $A$ by replacing every occurrence of the form $X^n t_1 t_2 ... t_n$ in $A$, where $X^n$ is free in $A$ and the $t_i$'s are pseudo-terms, by the pseudo-formula $S_{t_1 t_2}^{a_1 a_2} ... {}_{t_n}^{a_n} C$ and (ii) for every occurrence of the form $X^n t_1 t_2 ... t_n$ in $A$, where $X^n$ is free in $A$ and the $t_i$'s are pseudo-terms, it holds that if $x$ occurs in $t_i$ ($i \leqslant n$), then no occurrence of $a_i$ in $C$ belongs to the scope of an occurrence in $C$ of a quantifier that is immediately followed by $x$ (in other words, no pseudo-formula $S_{t_1 t_2}^{a_1 a_2} ... {}_{t_n}^{a_n} C$ considered in (i) contains a bound variable occurrence that does not already belong to $C$).

On the use of the $\forall^2 I$- and $\exists^2 E$-rules, there are restrictions as in the case of individual variables. In the terminology of Chapter I, $\forall^2 I$ and $\exists^2 E$ are thus improper inference rules for which there are corresponding *deduction rules*, obtained from the deduction rules for $\forall I$ and $\exists E$ by changing individual parameters and variables to predicate parameters and variables with the same number of argument places.

*Version 2.* The rules for $\forall^2 I$ and $\exists^2 E$ are as in version 1. The rules for $\forall^2 E$ and $\exists^2 I$, which now parallel more closely the corresponding rules for individual variables, are indicated by the figures:

$$\forall^2 E) \quad \frac{\forall X^n A}{A_T^{X^n}} \qquad\qquad \exists^2 I) \quad \frac{A_T^{X^n}}{\exists X^n A}$$

The two additional rules for $\lambda$ are indicated by the figures:

$$\lambda\text{I)} \quad \frac{S^{x_1\,x_2\,\cdots\,x_n}_{t_1\,t_2\,\cdots\,t_n} A}{\{\lambda x_1 x_2 \ldots x_n A\}\, t_1\, t_2 \ldots t_n} \qquad\qquad \lambda\text{E)} \quad \frac{\{\lambda x_1 x_2 \ldots x_n A\}\, t_1\, t_2 \ldots t_n}{S^{x_1\,x_2\,\cdots\,x_n}_{t_1\,t_2\,\cdots\,t_n} A}$$

INFERENCE RULES FOR RAMIFIED 2ND ORDER LOGIC. The inference rules for ramified second order logic are like those for simple second order logic except that we add certain restrictions concerning levels. A formula (different from $\wedge$) or predicate term is said to be of level $m$ if it (i) contains either some predicate variable of level $m$-1 or some predicate parameter or constant of level $m$, but (ii) contains no predicate variable of higher level than $m-1$ and no predicate parameter or constant of higher level than $m$. $\wedge$ is of level 1. The added restrictions are then as follows: To the above formulation of the $\forall^2$I- and $\exists^2$E-rules, we add that $P^n$ and $X^n$ are to be of the same level. To the above formulation of the $\forall^2$E- and $\exists^2$I-rules, we add in case of version 1 [version 2] that the pseudo-formula [term] that is substituted for $X^n$ (i.e. the $C$ in the definition of $A^*$) [(i.e. $T^n$)] is of the same or lower level than that of $X^n$.

SYSTEMS OF NATURAL DEDUCTION FOR 2ND ORDER LOGIC. By the addition of the proper inference rules and deduction rules of the various kinds considered above to the Gentzen-type systems for classical, intuitionistic, and minimal predicate logic of first order (stated in Chapter I), we get what I shall call (version 1 and 2 of) the Gentzen-type systems for *classical, intuitionistic, and minimal simple [ramified] predicate logic of second order*.[1] The systems are denoted $C^2$, $I^2$ and $M^2$ [$C_r^2$, $I_r^2$, and $M_r^2$]. The first version of $C^2$ and $C_r^2$ can easily be seen to be equivalent to e.g. the calculi for the respective logics stated by Church [2]. The first and second versions of the respective systems are equivalent in the following sense: Let $A^*$ be obtained from $A$ by successively replacing each part of the form $\{\lambda x_1 x_2 \ldots x_n B\} t_1 t_2 .. t_n$ by $S^{x_1\,x_2\,\cdots\,x_n}_{t_1\,t_2\,\cdots\,t_n} B'$, where $B'$ is obtained from $B$ by alphabetic changes of variables to avoid conflict with the variables occurring in any of the pseudo-terms $t_1, t_2, ..., t_n$. Let $\Gamma^*$ be obtained from $\Gamma$ by the same transformation with respect to each formula in $\Gamma$. Then $\Gamma \vdash A$ holds in the second version if and only if $\Gamma^* \vdash A^*$ holds in the first version. — The

---

[1] From a strictly intuitionistic view-point, probably only ramified predicate logic is meaningful.

second version is in many ways advantageous. Besides the simplification of the statement of the $\forall^2$E- and $\exists^2$I-rules,[1] there is the advantage —of particular value in the present connection—that we can state $\forall^2$- and $\exists^2$-reductions completely analogously to the definitions of the $\forall$- and $\exists$-reductions in Chapter II, § 2.

DEFINABILITY OF &, ∨, $\exists$ AND $\wedge$ IN INTUITIONISTIC LOGIC. In contrast to the situation for first order logic (cf. Corollary 9 in Chapter IV), we have:

THEOREM 1. *The logical constants* &, ∨, *and* $\exists$ *are definable in the systems* $I^2$, $M^2$, $I_r^2$ *and* $M_r^2$. *In particular it holds in these systems:*[2]

$$\vdash (A\,\&\,B) \equiv \forall X((A \supset (B \supset X)) \supset X)$$

$$\vdash (A \vee B) \equiv \forall X((A \supset X)\,\&\,(B \supset X) \supset X)$$

$$\vdash \exists x A \equiv \forall X(\forall x(A \supset X) \supset X)$$

$$\vdash \exists Y^n A \equiv \forall X(\forall Y^n(A \supset X) \supset X)$$

*where X is a* 0-*place predicate parameter, in the ramified case of the same level as* $A\,\&\,B$, $A \vee B$, $\exists x A$, *and* $\exists Y^n A$ *respectively. Furthermore,* $\wedge$ *is definable in* $I^2$ *and* $I_r^2$, *where it holds:* $\vdash \wedge \equiv \forall X X$ (*X is to be a* 0-*place parameter that in the ramified case may be of arbitrary level, e.g. the level* 1).

*Proof.* I prove the second and third formula by the four deductions below. The fourth formula is proved in the same way as the second. $a$ is to be a parameter that does not occur ~~not occur~~ in $A$.

[1] A different but related way of avoiding the inconvenience of the usual substitution rule for predicate variables is considered in Henkin [1].

[2] The definition of &, which is of interest also for classical logic, is due to Russell [1]. Also the definition of $\sim A$ as $A \supset \forall X X$ can be found there.

$A \vee B \quad as \quad \forall X((A \supset X) + (B \supset X) \supset X)$

$$\cfrac{\cfrac{\cfrac{\overset{(1)}{A}}{A \vee B}}{\overset{(1)}{A \supset (A \vee B)}} \quad \cfrac{\cfrac{\overset{(2)}{B}}{A \vee B}}{\overset{(2)}{B \supset (A \vee B)}}}{(A \supset A \vee B) \,\&\, (B \supset A \vee B)}$$

$$\cfrac{\forall X((A \supset X) \,\&\, (B \supset X) \supset X) \qquad (A \supset A \vee B) \,\&\, (B \supset A \vee B) \supset (A \vee B)}{A \vee B}$$

$$\cfrac{\cfrac{\cfrac{\cfrac{A_a^x \quad \cfrac{\overset{(2)}{\forall x(A \supset P)}}{A_a^x \supset P}}{P}}{\overset{(1)}{P}}}{\overset{(2)}{\forall x(A \supset P) \supset P}}}{\forall X(\forall x(A \supset X) \supset X)} \quad \exists xA$$

$$\cfrac{\cfrac{\cfrac{\cfrac{\overset{(1)}{A_a^x}}{\exists xA}}{A_a^x \supset \exists xA}}{\forall x(A \supset \exists xA)} \quad \cfrac{\forall X(\forall x(A \supset X) \supset X)}{\forall x(A \supset \exists xA) \supset \exists xA}}{\exists xA}$$

## § 2. Normal deductions in ramified 2nd order logic

Some of the results in Chapters II–IV can be extended to ramified second order logic. The terminology used in those chapters are extended to second order logic in an obvious way. We note that the lemma on parameters holds for second order logic also when extended to predicate variables.

REDUCTIONS. The inversion principle obviously holds also for the new inference rules, using the second version. As already remarked, $\forall^2$- and $\exists^2$-*reductions* are defined in complete analogy to the way we defined $\forall$- and $\exists$-reductions in Chapter II, § 2. To take $\forall^2$-reductions as an example, the deduction shown to the right below is a $\forall^2$-reduction of the deduction to the left (provided, as usual, that this deduction has only pure parameters):

$$\cfrac{\cfrac{\cfrac{\Sigma_1}{A}}{\forall X A_X^P}}{\cfrac{(A_{XT}^{PX})}{\Pi_2}} \qquad \cfrac{\cfrac{\Sigma_{1T}^P}{(A_{XT}^{PX})}}{\Pi_2}$$

Noting that $T$ is of the same or lower level than that of $P$, we can verify in the same way as for $\vee$-reductions that this reduction gives a new deduction of the same formula from the same set.

Furthermore, we need $\lambda$-*reductions*, trivially accomplished as shown below:

$$\frac{\dfrac{\Sigma_1}{S^{x_1 x_2 \ldots x_n}_{t_1 t_2 \ldots t_n} A}}{\dfrac{\{\lambda x_1 x_2 \ldots x_n A\} t_1 t_2 \ldots t_n}{(S^{x_1 x_2 \ldots x_n}_{t_1 t_2 \ldots t_n} A)}}{\Pi_2} \qquad \frac{\dfrac{\Sigma_1}{(S^{x_1 x_2 \ldots x_n}_{t_1 t_2 \ldots t_n} A)}}{\Pi_2}$$

A theorem on normal deductions can now be obtained by somewhat modifying the proofs for the first order. By a *normal deduction* we understand (almost as before) a deduction that contains no maximum segment and no redundant application of $\vee$E, $\exists$E, or $\exists^2$E. As in Chapter III, we confine the treatment of classical logic to the system $C_r^{2\prime}$, obtained from $C_r^2$ by omitting the $\vee$-, $\exists$-, and $\exists^2$-rules (and thus intended for languages not containing $\vee$ or $\exists$). In view of Theorem 1, the same reduction of $I_r^2$ and $M_r^2$ could now have been made.

THEOREM 2. *If $\Gamma \vdash A$ in the second version of one of the systems $C_r^{2\prime}$, $I_r^2$, and $M_r^2$, then there is a normal deduction in that system of $A$ from $\Gamma$.*[1]

*Proof.* Instead of using the degrees of the formulas, I shall now use their ranks. The *rank* of a formula $A$ is the pair $\langle m, q \rangle$ such that $m$ is the level of $A$ and $q$ is the number of occurrences of logical consatnts in $A$ with scopes containing a bound variable of level $m-1$ if $m>1$ and $q$ is the degree of $A$ if $m=1$. The rank of a segment is the rank of the formula that occurs in the segment. $\langle m', q' \rangle$ is *lower* than $\langle m, q \rangle$ if and only if $m' < m$ or both $m' = m$ and $q' < q$. Note that the new maximum segments that may arise through a $\forall^2$- or $\exists^2$-reduction are of lower rank than the removed one.

The proof may now proceed along the same lines as in Chapters III and IV. Substituting "rank" for "degree", the proof in the case of intuitionistic or minimal logic goes through without changes. It has only to be verified that the transformations really lower the induction value

---

[1] Theorem 1 in Chapter III can be established also for $C_r^{2\prime}$ (and for $C^{2\prime}$) but it does not simplify the proof of the present theorem, since a $\forall^2$-reduction may destroy the property that the consequences of the $\lambda_C$-rule are atomic.

also by the new definition. (When the length of the maximum segment $\sigma$ is greater than one, the last formula occurrence in $\sigma$ may now also be the consequence of $\exists^2 E$, but that case is treated in the same way as the $\exists E$-case.)

In the case of classical logic, it remains to consider the maximum segments that are consequences of the $\wedge_c$-rule. The complication then arises that an assumption discharged by an application of the $\wedge_c$-rule may be turned into a maximum segment by the transformations described in the proof of Theorem I in Ch. III, and such a new maximum formula obviously has a higher rank than the removed one. However, this is no serious difficulty, since we can at once remove all such maximum segments. How this is to be done is considered in more detail for the case when we have a deduction $\Pi$ of the form shown to the left below; the other cases being treated in the same general way. We transform it to the deduction $\Pi^*$ as shown to the right.

$$
\begin{array}{cc}
& \dfrac{\forall XA}{A_T^X \qquad \sim A_T^X} \\[2pt]
[\sim \forall XA] & \wedge \\[2pt]
\Sigma & [\sim \forall XA] \\[2pt]
\dfrac{}{\wedge} & \Sigma \\[2pt]
\forall XA & \wedge \\[2pt]
(A_T^X) & (A_T^X) \\[2pt]
\Pi_1 & \Pi_1
\end{array}
$$

$[\sim \forall XA]$ is to be the set of assumptions discharged by the application of the $\wedge_c$-rule in question. It will be assumed that no assumption in $[\sim \forall XA]$ is side-connected with a consequence of another application of the $\wedge_c$-rule; by obvious transformations this can always be arranged. We now consider a formula occurrence in $[\sim \forall XA]$ that is a maximum formula in $\Pi^*$. Such a formula occurrence stands within a part $\Pi'$ of $\Pi^*$ of the form

$$
\begin{array}{cc}
& \dfrac{\forall XA}{A_T^X \qquad \sim A_T^X} \\[2pt]
\Sigma' & \wedge \\[2pt]
\dfrac{\forall XA \qquad \sim \forall XA}{} \\[2pt]
\wedge
\end{array}
$$

If the end-formula of $(\Sigma'/\forall XA)$ is the consequence of an E-rule, then we replace $\Pi'$ by its $\supset$-reduction at $\sim\forall XA$. If the end-formula of $(\Sigma'/\forall XA)$ is the consequence of an I-rule, then it has the form $((\Sigma_1/A_P^X)/\forall XA)$ and we replace $\Pi'$ by

$$\frac{\dfrac{\Sigma_1{}_T^P}{A_{PT}^{XP} \qquad \sim A_T^X}}{\lambda}$$

In this way, all the maximum formulas in $[\sim\forall XA]$ in $\Pi$ can be removed without giving rise to any new maximum segments. We can thus obtain a deduction in which the number of maximum segments of the same rank as $\forall XA$ is lower than in $\Pi$, which was to be shown.

The theorems about the form of the normal deductions, i.e. Theorem 3 in Chapter III and Theorem 2 in Chapter IV, can be taken over if we leave out what is said about subformulas (since we do not define this notion for second order logic).

We note the following corollary:

COROLLARY 1. *If $\Pi$ is a normal deduction of $A$ from $\Gamma$ in $C_r^{2\prime}$, $I_r^2$, or $M_r^2$, and if $A$ and the formulas of $\Gamma$ contain no (quantifier with) predicate variables and no occurrence of $\lambda$, then $\Pi$ is a deduction also in the system $C'$, $I$, or $M$ respectively.*

*Proof.* Consider any path in $\Pi$. If a formula occurrence in the I-part [E-part] or in the minimum segment of the path contains a quantifier with a predicate variable, then obviously the first [last] formula occurrence in the path contains such a quantifier.

Among other corollaries, we note that the consistency of the systems in question is immediate from Corollary 1, and that Corollaries 6 and 7 in § 3 of Ch. IV fail in the general case (cf. Theorem 1) but hold for the special case when $\Gamma$ is empty.

## § 3. The unprovability by finitary methods of the theorem on normal deductions for simple 2nd order logic

Although the inversion principle also holds for simple second order logic (and corresponding reductions can be defined as in § 2), it is impossible to show by strictly finitary methods that all maximum formulas can be removed from a given deduction. By a result of Takeuti [1], which is proved below, a theorem of this kind implies the consistency of number theory (among others). Since the proof of this fact can be formalized within elementary number theory, it follows by the results of Gödel that the proof of the theorem on normal deductions cannot.[1]

A system for Peano arithmetic can be constructed in the following way. We consider the particular second order language that contains exactly three descriptive constants, namely the individual constant 0, the 1-place operation constant s (for the successor function) and the 2-place predicate constant =. An atomic formula consisting of = followed by two terms $t$ and $u$ will be denoted by $t=u$ or $(t=u)$.

We consider five particular *axioms* of the following form, where $x$, $y$ and $z$ are distinct and $X$ is a 1-place predicate variable:

$$(1) \quad \forall x(x=x) \qquad (2) \quad \forall x \forall y \forall z(x=z \ \& \ y=z \supset x=y)$$

$$(3) \quad \forall x \forall y(x=y \equiv sx=sy) \qquad (4) \quad \forall x \sim (0=sx)$$

$$(5) \quad \forall X(X0 \ \& \ \forall x(Xx \supset Xsx) \supset \forall x Xx)$$

The system $P^2$ [$P^{2*}$] is the system obtained by adding the axioms (1)–(5) [(1)–(4)] to $C^2$ (or, if we want, to $C^{2'}$ omitting the $\lor$-, $\exists$- and, $\exists^2$-rules). The system $P^{1*}$ is the system obtained by adding axioms (1)–(4) to $C$ (or to $C'$).

THEOREM 3. *The theorem on normal deductions for* $C^2$ (*or* $C^{2'}$) *implies the consistency of* $P^2$.

*Proof.* The theorem follows immediately from the three lemmata below (e.g., by showing that $0=s0$ is not provable in $P^2$ under the hypothesis of the theorem).

Let $F$ be a particular formula of the form

$$\forall X(X0 \ \& \ \forall x(Xx \supset Xsx) \supset Xa)$$

where $X$ is a 1-place predicate parameter. By the notations $Fx$, $Ft$ etc., I denote the result obtained from $F$ by replacing its parameter with $x$, $t$, etc. Let $A^F$ be the formula obtained from $A$ by relativizing the

---

[1] Löb [1] has recently announced a proof of Gentzen's Hauptsatz for higher order logic, using Gödel's set-theoretical axiom of constructibility ($V = L$).

*The announced proof was never carried out.*

quantifiers in $A$ that are followed by individual variables to $F$, i.e. by replacing every occurrence in $A$ of a pseudo-formula of the form $\forall xB$ [$\exists xB$] with $\forall x(Fx \supset B)$ [$\exists x(Fx \& B)$]. Similarly, let $\Gamma^F$ be obtained from $\Gamma$ by transforming every formula in $\Gamma$ in this way.

By induction on the length of the deductions in $P^2$, one easily shows

LEMMA 1. *Let $\Delta$ be the set of the formulas $Fa$ where $a$ occurs in $A$ or in some formula of $\Gamma$. Then:*

$$\Gamma \underset{P^2}{\vdash} A \quad implies \quad \Delta \cup \Gamma^F \underset{P^{2*}}{\vdash} A^F.$$

The following lemma would be an immediate corollary of a theorem on normal deductions for $C^2$ (or $C^{2'}$) in the same way as Corollary 1 in § 2:

LEMMA 2. *If $A$ contains no predicate variable and no occrrence of $\lambda$, then*

$$\underset{P^{2*}}{\vdash} A \quad implies \quad \underset{P^{1*}}{\vdash} A.$$

The consistency of $P^{1*}$ is established by the following easy corollary of Theorems 2 and 3 in Chapter III:

LEMMA 3. *If $\underset{P^{1*}}{\vdash} t=u$, then $t=u$.*

# VI. MODAL LOGIC

## § 1. Natural deduction for modal logic

By extending the systems of Chapter I through rules for modal concepts, we get Gentzen-type systems of natural deduction for modal logic based respectively on classical, intuitionistic, and minimal logic. I shall consider some alternative versions of modal rules, which when added to the classical system, give modal systems corresponding to what is known as S4 and S5 with quantifiers.

To the *languages* of Chapter I, we now add a new logical constant, the *modal operator* N (read: it is necessary that). In the definition of the formulas, we add a fourth clause:

4) If $A$ is a formula, so is N$A$.

The notion of principal sign is extended accordingly. A formula is said to be *modal* if its principal sign is N.

Also for N, we shall state an introduction and an elimination rule. In the first version that we shall consider, the introduction rule for N in the systems inspired by S4 allows us to infer N$A$ when we have either a proof of $A$ or a deduction of $A$ depending on modal formulas only. In the systems inspired by S5, N$A$ may be inferred when $A$ depends also on the negations of modal formulas. The elimination rule for N will allow us to infer $A$ from N$A$.

Using the terminology of Chapter I, we thus have an *improper inference rule* NI and a *proper inference rule* NE indicated by the figures:

$$\text{NI)} \quad \frac{A}{NA} \qquad \text{NE)} \quad \frac{NA}{A}$$

and a corresponding *deduction rule* for NI defined by the condition: An instance of the deduction rule for NI in S4 [S5] is to have the form $\langle\langle \Gamma, A\rangle, \langle \Gamma, NA\rangle\rangle$, where every formula of $\Gamma$ is modal [modal or the negation of a modal formula].

Thus, the Gentzen-type system of natural deduction for *classical*, *intuitionistic* and *minimal* $S4$ [$S5$] is got by adding the inference rule NE and the deduction rule for NI in $S4$ [$S5$] to the Gentzen-type system for classical, intuitionistic and minimal predicate logic respectively. The systems will be denoted $C_{S4}$, $I_{S4}$, $M_{S4}$, $C_{S5}$, $I_{S5}$, and $M_{S5}$ respectively. One easily shows the equivalence of $C_{S4}$ and $C_{S5}$ to usual systems of axiomatic type for $S4$ and $S5$.[1]

REMARK I. So-called *strict implication* ($\prec$) can be defined in the usual way, taking $A \prec B$ as a (syntactical) abbreviation for $N(A \supset B)$. Alternatively one can take $\prec$ instead of N as a primitive sign understanding $NA$ as short for $(A \prec A) \prec A$. A formula is said to be *strict* if its principal sign is $\prec$. We then have a deduction rule for $\prec I$ defined by the condition: An instance of the deduction rule for $\prec I$ in $S4$ [$S5$] is to have the form $\langle\langle \Gamma, B \rangle, \langle \Delta, A \prec B \rangle\rangle$, where $\Delta = \Gamma - \{A\}$ and all formulas of $\Delta$ are strict [strict or the negation of strict formulas]. Further, there is an elimination rule for $\prec$ similar to $\supset E$, i.e. instances of $\prec E$ have the form $(A, A \prec B/B)$. One easily shows the equivalence of the systems obtained in this way to the ones considered above.

REMARK 2. One may add an additional modal operator, $\diamond$ expressing *possibility*, and inference rules indicated by the figures:

$$\diamond I) \quad \frac{A}{\diamond A} \qquad \diamond E) \quad \frac{\diamond A \quad \overset{(A)}{B}}{B}$$

The way in which applications of $\diamond E$ discharge assumptions is expressed by the condition: An instance of the *deduction rule* for $\diamond E$

---

[1] Systems of axiomatic type for $S4$ and $S5$ (without quantifiers) are stated in Lewis–Langford [1]. A more elegant and transparent axiom system for $S4$ is given in Gödel [2]. (In particular, the rule NI in $S4$ for the special case when $\Gamma$ is empty is due to Gödel [2] and is proved to be a derived rule for Lewis–Langford's system by McKinsey–Tarski [1].) A system of natural deduction with the rule NI for $S4$ and the rule NE is given in Curry [1]. Other systems of natural deduction for modal logic are developed in Fitch [2].

Modal systems with quantifiers seem first to have been considered by Barcan [1] and Carnap [1]. See also Kripke [1].

A modal system based on intuitionistic logic is considered in Fitch [1] (cf. note 2 on p. 76). The systems $I_{S5}$ and $M_{S5}$ are of limited interest, since $S5$ is in a sense "classical" in its nature (a natural extension of these systems would be to allow in the deduction rule for NI that $\Gamma$ also contains formulas of the form $NB \supset NC$).

in S4 [S5] is to have the form $\langle\langle \Gamma_1, \Diamond A\rangle, \langle \Gamma_2, B\rangle, \langle \Delta, B\rangle\rangle$, where $\Delta = \Gamma_1 \cup (\Gamma_2 - \{A\})$, every formula of $\Gamma_2 - \{A\}$ has the form $\mathrm{N}C$ or $\sim\Diamond C$ [$\mathrm{N}C$, $\sim\Diamond C$, $\sim\mathrm{N}C$, or $\Diamond C$] for some $C$, and $B$ has the form $\Diamond C$ or $\sim\mathrm{N}C$ [$\Diamond C$, $\sim\mathrm{N}C$, $\sim\Diamond C$, or $\mathrm{N}C$] for some $C$. If these rules are added to the modal systems considered above, then the condition on $\Gamma$ in the definition of the deduction rule for NI is to be modified and is to be indentical to the condition on $\Gamma_2 - \{A\}$ in the definition of the deduction rule for $\Diamond$E above. In the classical systems, it is of course possible to take only one of the modal operators as primitive and to define the other,[1] since one can prove the laws $\Diamond A \equiv \sim\mathrm{N}\sim A$ and $\mathrm{N}A \equiv \sim\Diamond\sim A$ in these systems.[1] It can be shown (cf. Corollary 9 in Chapter IV) that the same does not hold in the case of intuitionistic and minimal logic.[2] Nevertheless, in what follows, I shall confine my attention to the necessity operator.

## § 2. Essentially modal formulas

Although the restriction on the deduction rule for NI in § 1 is natural and easily stated, it forces certain deductions to proceed in a roundabout way, as is illustrated in the following deduction of $\mathrm{N}(A\,\&\,B)$ from $\mathrm{N}A\,\&\,\mathrm{N}B$, which is correct in all the modal systems considered above:

$$
\begin{array}{ccc}
& {\scriptstyle(1)} & {\scriptstyle(2)} \\
& \dfrac{\mathrm{N}A}{A} & \dfrac{\mathrm{N}B}{B}
\end{array}
$$

$$\dfrac{A\,\&\,B}{\mathrm{N}(A\,\&\,B)}$$

$$
\dfrac{\mathrm{N}A\,\&\,\mathrm{N}B}{\dfrac{\mathrm{N}A\,\&\,\mathrm{N}B}{\dfrac{\mathrm{N}A}{\dfrac{}{\mathrm{N}B \supset \mathrm{N}(A\,\&\,B)}}}\quad \dfrac{\dfrac{\mathrm{N}B \supset \mathrm{N}(A\,\&\,B)}{\mathrm{N}A \supset (\mathrm{N}B \supset \mathrm{N}(A\,\&\,B))}^{\,(2)}}{}^{\,(1)}}
$$

$$\dfrac{\mathrm{N}B \supset \mathrm{N}(A\,\&\,B)}{\mathrm{N}(A\,\&\,B)}$$

---

[1] If $\Diamond$ is taken as the only primitive, then the rule for $\Diamond$ E should be modified so that the $B$ in the definition of $\Diamond$ E above is also allowed to be $\wedge$.

[2] The intuitionistic modal logic proposed by Fitch [1] has primitives for both necessity and possibility. However, his system is in certain respects considerably weaker than S4 (not having $\mathrm{N}A \supset \mathrm{NN}A$ as theorem) and is in other respects stronger (having $\forall x\mathrm{N}A \supset \mathrm{N}\forall xA$ as axiom). Also the modal logic proposed by Curry [1] and [2] has primitives for both necessity and possibility, but his possibility obeys rather unusual laws such as $\Diamond A\,\&\,\Diamond B \supset \Diamond(A\,\&\,B)$.

For this reason, I shall liberalize the restriction on the deduction rule for NI using the notion of an *essentially modal formula with respect to the system* S, which I define inductively for the various systems as follows.

For the case when S is $M_{S4}$:

1) $NA$ is essentially modal with respect to S.

2) If $A$ and $B$ are essentially modal with respect to S, then so are $A \& B$ and $A \vee B$.

3) If $A_t^x$ is essentially modal with respect to S, then so is $\exists x A$.

For the case when S is $I_{S4}$ or $C_{S4}$, we add to 1)–3):

4) $\wedge$ is essentially modal with respect to S.

For the case when S is $M_{S5}$ or $I_{S5}$, we add to 1)–4):

5) If $A$ is essentially modal with respect to S, then so is $\sim A$.

For the case when S is $C_{S5}$ we add clauses also for $\supset$ and $\forall$, or we may simply say that $A$ is essentially modal with respect to $C_{S5}$, when each occurrence of a predicate parameter or predicate constant in $A$ stands within the scope of an occurrence of N; $A$ is then also said to be *modally closed*.

We then get a second version of the modal systems considered above by modifying the deduction rule for NI. An instance of the deduction rule for NI in the *second version* of the system S is to have the form $\langle\langle \Gamma, A\rangle, \langle \Gamma, NA\rangle\rangle$, where every formula of $\Gamma$ is essentially modal with respect to S.

The equivalence between the first and second versions of respective systems follows from the following lemma:

LEMMA. *If $A$ is essentially modal with respect to* S, *then $A \supset NA$ is provable in* S *as first defined.*

The lemma is proved by induction on the degree of $A$. The case when the principal sign of $A$ is $\vee$ or $\exists$ is immediate. For the case when $A$ is a conjunction, compare the deduction above. The case when the principal sign of $A$ is $\supset$ or $\forall$ arises only in $C_{S5}$, except for the trivial case in $I_{S5}$ and $M_{S5}$ when $A$ is of the form $\sim B$. We prove $(NA \supset NB) \supset N(NA \supset NB)$ in $C_{S5}$, using the $\wedge_c$-rule. It remains to consider the case of universal quantification for $C_{S5}$. $N\forall x A$ can be deduced from $\sim N \sim \forall x NA$ as follows, where it is assumed that $a$ is a parameter that does not occur in $A$:

$$
\begin{array}{c}
\dfrac{
\dfrac{
\dfrac{
\dfrac{\overset{(1)}{\forall x NA}}{NA_a^x} \qquad \overset{(2)}{\sim NA_a^x}
}{\Large\curlywedge}
}{\sim \forall x NA}
\quad (1)
}{N \sim \forall x NA \qquad\qquad \sim N \sim \forall x NA}
}{\dfrac{\dfrac{\dfrac{\dfrac{\Large\curlywedge}{NA_a^x} \;(2)}{A_a^x}}{\forall x A}}{N \forall x A}}
\end{array}
$$

Since $\forall x NA \vdash \sim N \sim \forall x NA$ holds trivially (cf. Remark 2 in § 1), this shows that $\forall x NA \vdash_{C_{S5}} N\forall x A$.[1]

Although the second version of the deduction rule for NI considerably simplifies many deductions, it is still not always possible to avoid maximum formulas. For a counter example, consider the following deduction of $NNP$ from $NP \& Q$:

$$
\dfrac{
\dfrac{NP \& Q}{NP} \qquad \dfrac{\dfrac{\overset{(1)}{NP}}{NNP}}{NP \supset NNP}\;(1)
}{NNP}
$$

I shall therefore further liberalize the restriction on the deduction rule for NI in the next section.

## § 3. Normal deductions in the S4-systems

In this section, I shall use the alternative way of defining the notion of deduction described in § 4 of Chapter I.

---

[1] This is a somewhat simplified proof of the so-called Barcan-formula, which is proved by Prior [1]. It was originally proposed as an axiom by Barcan [1] (at the suggestion of Fitch, cf. Fitch [1]) in a system which is much weaker than S5.

For the *third versions* of the various modal systems considered above, we then define:

$\langle \Pi, \mathcal{J} \rangle$ is a *deduction* in S if and only if $\Pi$ is a quasi-deduction[1] in S and $\mathcal{J}$ is a discharge-function for $\Pi$ satisfying the conditions (i) and (ii) on a regular discharge-function given in Ch. I, § 4, and the condition:

(iii) if $B$ is a premiss in $\Pi$ of an application of NI and $B$ depends with respect to $\mathcal{J}$ on $A$, then there is a formula occurrence $C$ below or identical to $A$ and above or identical to $B$ that fulfills the following three conditions:[2]

1) $C$ does not depend with respect to $\mathcal{J}$ on any assumptions that $B$ does not depend on with respect to $\mathcal{J}$;

2) $C$ does not contain any occurrence of a proper parameter of an application of $\forall$I [$\exists$E] whose premiss [minor premiss] stands above $B$ or is equal to $B$;

3) $C$ is essentially modal with respect to S.

The main idea in showing that the second and third versions of respective systems are equivalent can be indicated roughly as follows. Let $\langle \Pi, \mathcal{J} \rangle$ be a deduction, where $\Pi$ has the form $((\Sigma/B)/\mathrm{N}B/\Pi')$. Let $\Gamma = \{A_1, A_2, ..., A_n\}$ be the set of assumptions on which the end-formula of $(\Sigma/B)$ depends with respect to $\mathcal{J}$, and let $\Gamma^* = \{A_1^*, A_2^*, ... A_n^*\}$ be a set such that $A_i^*$ $(i \leqslant n)$ is a formula occurrence in $(\Sigma/B)$ that belongs to the thread which starts with $A_i$ and that satisfies conditions 1–3 in clause (iii) in the definition above. Let $\Sigma^*$ be obtained from $\Sigma$ by cutting off the parts of $\Sigma$ that stand above a formula occurrence that belongs to $\Gamma^*$. $((\Sigma^*/B)/\mathrm{N}B)$ is then a deduction of $\mathrm{N}B$ from $\Gamma^*$ (note condition 2 in clause (iii)) and the application of NI in question now satisfies the requirements in version 2. Let therefore $\mathrm{N}B$ be followed by a series of applications of $\supset$I yielding $(A_1^* \supset (A_2^* \supset ... (A_n^* \supset \mathrm{N}B)...))$. Now, if $\Sigma_i^*$ is the part of $\Sigma$ ending with $A_i^*$, then $\Sigma_i^*$ is obviously a deduction of $A_i^*$ from $\Gamma$ (note condition 1 in clause (iii)). By the use of these deductions and a series of applications of $\supset$E, we hence obtain an occurrence of $\mathrm{N}B$ that depends on the assumptions in $\Gamma$ only.

For the third version of the S4-systems, it can easily be verified that all the reductions defined in Chapter II, § 2 (see especially the Remark

---

[1] I.e., besides the rules of Chapter I, the improper inference rule NI and the rule NE may have been applied in $\Pi$.

[2] This restriction on NI is in certain respects similar to the one used by Fitch [2].

at the end of the section) yield corresponding deductions of the same kind. To these reductions we add N-*reductions* defined in the obvious way. By the same induction as in the proofs of Theorem 2 in Chapter III and Theorem 1 in Chapter IV, we can then prove the theorem below; in the classical case, we confine ourselves to the system $C'_{S4}$ (omitting the v- and ∃-rules) and remove maximum segments that are consequences of the $\wedge_c$-rule in the way described in the proof of Theorem 2 in Chapter V.

THEOREM 1. *If* $\Gamma \vdash A$ *in one of the systems* $C'_{S4}$, $I_{S4}$, *and* $M_{S4}$, *then there is a normal deduction of* $A$ *from* $\Gamma$ *in the third version of this system.*

The theorems about the form of normal deductions and most of the corollaries in Chapters III and IV can now be carried over to the S4-systems.

## § 4. Normal deductions in $C'_{S5}$

The argument in § 3 does not suffice to establish the theorem on normal deductions also for the S5-systems. To do so for the system $C_{S5}$, I shall exploit the analogy between N and the universal quantifier. Parallel to the more liberal version of the restriction on the ∀I-rule suggested in Chapter I (p. 28), we get a fourth version of $C_{S5}$ by replacing condition (iii) in the definition of the deductions in § 3 with

(iii') If $B$ is premiss of an application of NI and depends with respect to $\mathcal{J}$ on the assumption $A$, then every connection between $A$ and $B$ in the subtree of $\Pi$ determined by $B$ contains a formula occurrence that is modally closed.

Condition (iii) in § 3 is a special case of (iii'). Conversely, it can be shown by induction over the lengths of the deductions that the fourth version of $C_{S5}$ is not stronger than the third version. In the same way as in § 3, we now get

THEOREM 2. *If If* $\Gamma \vdash_{C'_{S5}} A$, *then there is a normal deduction of* $A$ *from* $\Gamma$ *in the fourth version of* $C'_{S5}$.

# VII. SOME OTHER CONCEPTS OF IMPLICATION

## § 1. Relevant implication

In both classical and intuitionistic logic, the logical constant $\supset$ corresponds to a rather weak meaning of the phrase "if ..., then ...". For instance, the formulas

(1) $$P \supset (Q \supset P) \quad \text{and} \quad \sim P \supset (P \supset Q)$$

are provable in both C and I. In minimal logic the first formula is provable but not the second (however $\sim P \supset (P \supset \sim Q)$ is provable in M.). If $\supset$ is replaced by strict implication (see Remark 1 in Chapter VI, § 1), then the formulas in (1) fail to be provable. But if, at the same time, we replace $P$ by $NP$ in the first formula and replace $\sim P$ by $N \sim P$ in the second formula, then they again become provable in $C_{S4}$ and in $I_{S4}$. (For $M_{S4}$ the analogue to what was said above holds.) Thus, even when $\supset$ is replaced by strict implication, the formulas

(2) $$Q \supset (P \supset P) \quad \text{and} \quad \sim (P \supset P) \supset Q$$

are provable in the systems $C_{S4}$ and $I_{S4}$.

One may therefore ask if one cannot formalize a stronger meaning of "if ..., then ..." according to which a necessary condition for "if $A$, then $B$" to hold is that $A$ is in some sense relevant to $B$. A system (called the weak implicational calculus with negation) that formalizes such a concept of implication has been introduced by Church [1]. In this system, whose logical constants consist of a sentential constant for falsehood ($\curlywedge$) and a binary sentential connective ($\supset$), none of the formulas in (1) and (2) are provable. A deduction theorem can be established for the system in the following weaker form. Given a deduction $\mathcal{D}$ of $B$ from $\{A_1, A_2, ..., A_n\}$ it holds: If $A_n$ is *relevantly used in deducing $B$ in $\mathcal{D}$*, then

$$A_1, A_2, ..., A_{n-1} \vdash A_n \supset B;$$

otherwise,

$$A_1, A_2, ..., A_{n-1} \vdash B.$$

The italicized phrase can be made precise in a natural way.

Church's system is of axiomatic type. A corresponding system of natural deduction is constructed by Anderson-Belnap [2] using a type of natural induction introduced by Fitch (see Appendix C), which is supplemented by a device of indices in order to keep track of the assumptions that have been relevantly used in deducing a certain formula occurrence.

A corresponding Gentzen-type system is obtained in a simpler way by simply modifying the $\supset$ I-rule so that $A \supset B$ may be inferred from $B$ only when $B$ depends on $A$ in the deduction in question; in other words, we require that some assumption is discharged by every application of $\supset$ I.[1] Notice that in the Gentzen-type systems, every assumption in a deduction is actually used in the deduction and is connected with the end-formula by a chain of applications of inference rules. If $\supset$ and $\wedge$ are the only logical constants that are present in a deduction, then every assumption in the deduction is, in the same sense as above, also relevantly used in deducing the end-formula.

More precisely, we may define the Gentzen-type system in question as follows. We consider *languages* that are as in Chapter I except for containing no logical constants besides $\supset$ and $\wedge$. We modify the *deduction rule* for $\supset$ I given in Chapter I so that $\langle\langle \varGamma, B \rangle, \langle \varDelta, A \supset B \rangle\rangle$ is now an instance of this deduction rule if and only if $A \in \varGamma$ and $\varDelta = \varGamma - \{A\}$. (Recall that in a deduction $\Pi$ of $A$ *depending on $\varGamma$*—rather than *from $\varGamma$*—every formula in $\varGamma$ actually occurs in $\Pi$ as an assumption). We then define what I shall call the *pure system for relevant implication* (denoted $\mathsf{R_p}$) as the system that contains exactly two rules, namely the proper inference rule for $\supset$ E and the deduction rule for $\supset$ I as defined above. One easily verifies the equivalence of this system to Church's system mentioned above.

Maximum formulas cannot always be avoided in deductions in $\mathsf{R_p}$ as now defined. As a counter-example, there is a deduction $\Pi$ of $A \supset (A \supset C)$ from $\{B, \ A \supset (A \supset (B \supset C))\}$ as below and it can be shown that there is no corresponding normal one:

---

[1] By comparison, recall that the same restriction on $\vee$E and $\exists$E, which is made in normal deductions, does not affect deducibility.

$$\frac{\begin{array}{c}\overset{(1)}{A} \quad \overset{(2)}{A \supset (B \supset C)}\end{array}}{\begin{array}{c}B \qquad\qquad B \supset C\end{array}}$$

$$\cfrac{\overset{(3)}{A} \quad A \supset (A \supset (B \supset C)) \qquad \cfrac{\cfrac{C}{A \supset C}^{(1)}}{(A \supset (B \supset C)) \supset (A \supset C)}^{(2)}}{\cfrac{A \supset (B \supset C)}{\cfrac{A \supset C}{A \supset (A \supset C)}^{(3)}}}$$

Note that the formula occurrence $(A \supset (B \supset C)) \supset (A \supset C)$ is a maximum segment in $\Pi$, and that the $\supset$-reduction $\Pi'$ of $\Pi$ with respect to this formula occurrence is no longer a correct deduction any more; the assumption (3) will be discharged in $\Pi'$ by the same application of $\supset I$ as the assumption (1), and this makes the last application of $\supset I$ incorrect according to the new restriction on this rule.

This difficulty vanishes if we instead use the alternative definition of deduction described in § 4 of Chapter I. For this reason, I shall henceforth understand by a deduction in $R_p$ a pair $\langle \Pi, \mathcal{J} \rangle$ such that $\Pi$ is a quasi-deduction obtained by the use of only $\supset I$ and $\supset E$ and $\mathcal{J}$ is a discharge-function for $\Pi$ that takes as values *exactly* the premisses of applications of $\supset I$.

It can now be seen that if $\Pi$ is a deduction in $R_p$, then a $\supset$-reduction of $\Pi$ as defined in Chapter II, § 2 (see especially the Remark at the end of the section) is a new correct deduction in $R_p$ of the same formula depending on the same set. We thus obtain

THEOREM 1. *If $\Gamma \vdash_{R_p} A$, then there is a normal deduction in $R_p$ of $A$ from $\Gamma$.*

Theorem 3 in Chapter III about the form of the normal deductions holds now also. Among the corollaries, we note the subformula principle, and the fact that none of the formulas in (1) and (2) above are provable in $R_p$.[1]

## § 2. Relevant implication extended by minimal and classical logic

If we want to consider a system that formalizes a stronger meaning of "if ..., then ..." as in § 1 but also contains the logical constants &,

---

[1] Church [1] shows by the matrix method that $P \supset (Q \supset P)$ is not provable in $R_P$.

v, V, and ∃ in their usual meaning, then the situation is more problematic. For instance, in the following two proofs of $P \supset (Q \supset P)$, every application of $\supset$ I satisfies the requirement of discharging some assumption:

$$
\begin{array}{cc}
(1) & (2) \\
P & Q \\
\hline
\multicolumn{2}{c}{P \,\&\, Q} \\
\hline
\multicolumn{2}{c}{P} \quad (1) \\
\hline
\multicolumn{2}{c}{Q \supset P} \quad (2) \\
\hline
\multicolumn{2}{c}{P \supset (Q \supset P)}
\end{array}
\qquad
\begin{array}{ccc}
(1) & & (3) \quad (4) \\
P & (2) & Q \quad Q \supset P \\
\hline
P \vee (Q \supset P) & P & P \quad (2),(4) \\
\hline
\multicolumn{3}{c}{P} \quad (3) \\
\hline
\multicolumn{3}{c}{Q \supset P} \quad (1) \\
\hline
\multicolumn{3}{c}{P \supset (Q \supset P)}
\end{array}
$$

Hence, when & or v is present, it is not sufficient that $B$ depends on $A$ for $A$ to be relevantly used in deducing $B$.[1] We have thus to reconsider the question of making the idea that "$A$ is relevantly used in deducing $B$" precise.[2]

The solution that I shall adopt here can be roughly indicated by saying that $A$ will not be considered to be relevantly used in deducing $B \,\&\, C$ in the deduction

$$
\begin{array}{cc}
\Sigma_1 & \Sigma_2 \\
B & C \\
\hline
\multicolumn{2}{c}{B \,\&\, C}
\end{array}
$$

unless it holds *both* that $A$ is relevantly used in deducing $B$ in ($\Sigma_1/B$) *and* that $A$ is relevantly used in deducing $C$ in ($\Sigma_2/C$). From this suggestion and dual considerations concerning v, one is led to the two following systems, which I shall call the (Gentzen-type) systems for *relevant implication extended with* respectively *minimal logic* and *classical logic*; they will be denoted $R_M$ and $R_C$ respectively. More precisely, the systems are described using the notions in § 4 of Chapter I as follows.

---

[1] Church [1] draws the conclusion that & and V with their usual properties cannot be adjoined to relevant implication. However, as shown below, it is possible to add the usual rules for & and V in such a way that deducibility among formulas not containing $\supset$ remains as in ordinary logic.

[2] We may note that the deductions above are not normal, and that in a normal deduction, all assumptions are, in a sense, relevantly used. In view of that, one may consider the system obtained from M by restricting the deductions to normal ones in which $\supset$ I is as in § 1. However, such a system has the disadvantages discussed in the Remark of Appendix B, § 1.

The *languages* that the systems are intended for are as in Chapter I.

The *improper inference rules* of $R_M$ [$R_C$] are the same as those of M [C], and hence, the notion of a quasi-deduction for $R_M$ [$R_C$] coincides with that for M [C].

Let us say that a formula occurrence $B$ in a quasi-deduction $\Pi$ *originates from* the formula occurrence $A$ in $\Pi$ if there is a sequence $A_1, A_2, ..., A_n$ of formula occurrences in $\Pi$ such that (cf. the definition of path) (i) $A_1 = A$; (ii) $A_1$ is an assumption not discharged by an application of $\vee$ E or $\exists$ E; (iii) $A_{i+1}$ ($i < n$) either stands immediately below $A_i$ or is an assumption discharged by an application of $\vee$ E or $\exists$ E whose major premiss is $A_i$; and (iv) $A_n = B$.

A *deduction* in $R_C$ is then defined as a pair $\langle \Pi, \mathcal{J} \rangle$ such that $\Pi$ is a quasi-deduction in $R_C$ and $\mathcal{J}$ is a discharge-function for $\Pi$ satisfying the conditions (i) and (ii) on a regular discharge-function (p. 30) and in addition, the following two conditions numbered (iv) and (v):

(iv) Every premiss [minor premiss] of an application of $\supset$ I or the $\wedge_C$-rule [$\vee$ E or $\exists$ E] is assigned some value by $\mathcal{J}$.

(v) If $\mathcal{J}(A^1) = B$ is a premiss of an application of $\supset$ I or the $\wedge_C$-rule and stands below a premiss [minor premiss] of an application $\alpha$ of &I [$\vee$ E] that originates from $A^1$ in the subtree $\Pi'$ of $\Pi$ determined by $B$, then the other premiss [minor premiss] of $\alpha$ originates in $\Pi'$ from some formula occurrence $A^2$ of the same shape as $A^1$ such that also $\mathcal{J}(A^2) = B$.

The deductions in $R_M$ are defined in the same way except that we leave out what is said about the $\wedge_C$-rule.

It can again be seen that if $\Pi$ is a deduction of $A$ from $\Gamma$ in $R_M$ or $R_C$, then a reduction of $\Pi$ as defined in Chapter II, § 2 is also a deduction of $A$ from $\Gamma$ in $R_M$ or $R_C$ respectively. Let $R_{C'}$, be the system obtained from $R_C$ by leaving out the rules for $\vee$ and $\exists$. We can then prove in the same way we proved the corresponding theorems in Chapters III and IV:

THEOREM 2. *If $\Gamma \vdash A$ in $R_M$ or $R_{C'}$, then there is a normal deduction in that system of $A$ from $\Gamma$.*

The theorems in Chapter III and IV about the form of the normal deductions hold now also. Among the corollaries, we note especially the separation theorem; hence, deducibility in $R_M$ among formulas not containing $\supset$ is as in M, and deducibility in $R_M$ among formulas containing no other logical constants besides $\supset$ and $\wedge$ is as in $R_p$.

However, it turns out that the system $R_{C'}$ is not equivalent to $R_C$. Although $A \vee B \equiv \sim(\sim A \,\&\sim B)$ and $\exists x A \equiv \sim \forall x \sim A$ are provable in $R_C$, the $\vee$- and $\exists$-rules do not become derived rules in $R_{C'}$ when $\vee$ and $\exists$ are defined in the corresponding way. And in $R_C$ one cannot always avoid maximum segments that are consequences of the $\wedge_C$-rule and major premisses of $\vee E$ (unlike the situation for $C$, where this can be shown to be possible). The results for $R_{C'}$ are therefore of limited interest.

## § 3. Rigorous implication

If we add modal rules for N to the systems for relevant implication, we can define a still stronger concept of implication, which comes close to the *rigorous implication* (German: *strenge Implikation*) introduced by Ackermann [1] and studied especially by Anderson-Belnap.

I shall consider the two systems that arise by adding the modal rules in S4 to the systems $R_M$ and $R_C$; they will be denoted $R_{M,S4}$ and $R_{C,S4}$. In more detail they are described as follows.

The *languages* that we consider are as in Chapter I with the addition of N. We use the third version of S4 (Chapter VI, § 3), and define a *deduction* in $R_{M,S4}$ [$R_{C,S4}$] as a pair $\langle \Pi, \mathcal{J} \rangle$ such that $\Pi$ is a quasi-deduction in $M_{S4}$ [$C_{S4}$] and $\mathcal{J}$ is a discharge-function for $\Pi$ satisfying the conditions (i) and (ii) on a regular discharge-function (p. 30), condition (iii) in the definition of deduction in the third version of $M_{S4}$ [$C_{S4}$] (p. 79), and conditions (iv) and (v) in the definition of deduction in $R_M$ [$R_C$] (p. 85).

Let $A \prec B$ be an abbreviation for $N(A \supset B)$ (cf. Remark 1, in Chapter VI, § 1); $\prec$ now stands for *rigorous implication*. Consider for the moment only those formulas in which every occurrence of an implication (i.e., of a pseudo-formula of the form $(B \supset C)$) is immediately preceded by N. It is conjectured that the sentential part of the sub-system of $R_{C,S4}$ that is obtained in this way is equivalent to the so-called calculus of entailment (E), a modification of Ackermann's system for rigorous implication proposed by Anderson-Belnap [1] (see e.g. one of the formulations of E found in Anderson [1]), and that the corresponding sub-system of $R_{M,S4}$ is equivalent to the system obtained from E by dropping the axioms of the form $\sim \sim A \prec A$.

By the same arguments that have been used for the previous systems, we have a *theorem on normal deductions* for $R_{M,S4}$ and for $R_{C',S4}$

(concerning the relations between $R_{C,S4}$ and $R_{C',S4}$—the latter being the system obtained from the former by omitting the v- and ∃-rules— the same holds as was said for $R_C$ and $R_{C'}$). This solves some problems for $R_{M,S4}$ that have been raised for E by Anderson-Belnap (see e.g. Anderson [2]). As a corollary of the theorem on normal deductions, we have the separation theorem as usual (cf. Corollary 4 in Chapter IV). This means e.g. that the formulas provable in the pure calculus of entailment ($E_I$, see Anderson-Belnap [2]) coincide with the formulas that are provable in $R_{M,S4}$ and are formulated with no other logical constant than that for rigorous implication.

Furthermore, we have as a corollary for $R_{M,S4}$: if $\vdash A$ and $\vdash \sim A \vee B$, then $\vdash B$. Because, from $\vdash \sim A \vee B$ it follows that either $\vdash \sim A$ or $\vdash B$ (see Corollary 6, Chapter IV), but the first possibility is excluded if also $\vdash A$ (Corollary 5, Chapter IV).[1] This means that the system E and Ackermann's system for rigorous implication are equivalent when the axioms of the form $\sim \sim A \prec A$ are dropped from both systems, provided that the equivalence conjectured above holds, i.e. the equivalence between the described sub-system of $R_{M,S4}$ and the system obtained from E by dropping the axioms of the form $\sim \sim A \prec A$.[2] (Whether the same holds when these axioms are not omitted is not known.)

---

[1] This holds also for M by the same argument (as first shown by Johansson [1]). But note that $(A \& (\sim A \vee B)) \supset B$ is not generally provable in M.

[2] The equivalence is meant as equivalence with respect to provability. Furthermore, E is to be understood as containing appropriate rules for V since it is not possible to define V when the axioms $\sim \sim A \prec A$ are dropped.

# APPENDIX A

## THE CALCULI OF SEQUENTS

## § 1. Definition of the calculi of sequents

Calculi of sequents for classical, intuitionistic and minimal logic can be formulated as follows.[1]

The *languages* considered are the same as in Chapter I with the addition of one new non-descriptive symbol, denoted: $\rightarrow$. A symbol string of the form $\varnothing \rightarrow \Psi$, where $\varnothing$ and $\Psi$ are (possibly empty) strings of formulas, is called a *sequent*; $\varnothing$ is its *antecedent* and $\Psi$ its *succedent*. However, in the calculi for intuitionistic and minimal logic, we consider only sequents where the succedent consists of at most one formula.[2]

I shall use the notation $\varGamma \rightarrow \varDelta$ to denote a sequent whose antecedent [succedent] is a string of the formulas in $\varGamma$ [$\varDelta$]; for definiteness, we may fix a certain alphabetic order of the formulas and decide that the antecedent [succedent] is the string of the formulas in $\varGamma$ [$\varDelta$] taken in that order (without repetitions). Notations such as $A,\ \varGamma \rightarrow \varDelta$ and $\varGamma \rightarrow \varDelta, A$ are used to denote the sequents $\{A\} \cup \varGamma \rightarrow \varDelta$ and $\varGamma \rightarrow \varDelta \cup \{A\}$ respectively.

---

[1] Calculi of sequents for classical and intuitionistic logic were developed by Gentzen [3]. The calculi presented here are slight modifications of Gentzen's calculi. A calculus of sequents for minimal logic (obtained from the one for intuitionistic logic by omitting one rule) was stated by Johansson [1]. A survey of some variants of calculi of sequents is given in Curry [3], where also several references can be found. Some other variants of the calculi of sequents have been constructed by Schütte. In Schütte [1], a classical sequent $\varGamma \rightarrow \varDelta$ is replaced by a disjunction of the negations of the formulas in $\varGamma$ and the formulas in $\varDelta$, and an intuitionistic sequent $A_1 A_2 \ldots A_n \rightarrow B$ is replaced by the formula $A_1 \supset (A_2 \supset \ldots \supset (A_n \supset B)\ldots)$; the rules are otherwise as in Gentzen's calculi. In Schütte [2], the inference rules are generalized so that they allow transformations also of various subformulas of a given formula.

[2] Systems of sequents with just one formula in the succedent were already introduced by Hertz [1] and studied by Gentzen [1].

A *proof* in one of the calculi considered here is a tree of sequents in which (i) every top-sequent is of the form $\Gamma \rightarrow \Delta$ where $\Gamma$ and $\Delta$ have a common element; (ii) every other sequent occurrence $S$ stands immediately below one or two sequent occurrences, say $S_1$, or $S_1$ and $S_2$, such that $(S_1/S)$ or $(S_1, S_2/S)$ respectively is an instance of one of the inference rules listed below, excluding the $\wedge$-rule in case of minimal logic.

A proof is said to be a proof *of* the sequent that occurs as end-sequent. A sequent is *provable* if there is a proof of it.

An *instance* of an *inference rule* is to be of one of the forms indicated below with the *restriction*: the parameter $a$ is not to occur in the consequence of an instance of the rule $\rightarrow \forall$ or $\exists \rightarrow$.

$$\rightarrow \&) \quad \frac{\Gamma \rightarrow \Delta, A \quad \Gamma \rightarrow \Delta, B}{\Gamma \rightarrow \Delta, A \& B} \qquad \& \rightarrow) \quad \frac{A, \Gamma \rightarrow \Delta}{A \& B, \Gamma \rightarrow \Delta} \quad \frac{B, \Gamma \rightarrow \Delta}{A \& B, \Gamma \rightarrow \Delta}$$

$$\rightarrow \vee) \quad \frac{\Gamma \rightarrow \Delta, A}{\Gamma \rightarrow \Delta, A \vee B} \quad \frac{\Gamma \rightarrow \Delta, B}{\Gamma \rightarrow \Delta, A \vee B} \qquad \vee \rightarrow) \quad \frac{A, \Gamma \rightarrow \Delta \quad B, \Gamma \rightarrow \Delta}{A \vee B, \Gamma \rightarrow \Delta}$$

$$\rightarrow \supset) \quad \frac{A, \Gamma \rightarrow \Delta, B}{\Gamma \rightarrow \Delta, A \supset B} \qquad \supset \rightarrow) \quad \frac{\Gamma \rightarrow \Delta_1, A \quad B, \Gamma \rightarrow \Delta_2}{A \supset B, \Gamma \rightarrow \Delta_1 \cup \Delta_2} \qquad \wedge) \quad \frac{\Gamma \rightarrow \Delta, \wedge}{\Gamma \rightarrow \Delta, A}$$

$$\rightarrow \forall) \quad \frac{\Gamma \rightarrow \Delta, A}{\Gamma \rightarrow \Delta, \forall x A_x^a} \qquad \forall \rightarrow) \quad \frac{A_t^x, \Gamma \rightarrow \Delta}{\forall x A, \Gamma \rightarrow \Delta}$$

$$\rightarrow \exists) \quad \frac{\Gamma \rightarrow \Delta, A_t^x}{\Gamma \rightarrow \Delta, \exists x A} \qquad \exists \rightarrow) \quad \frac{A, \Gamma \rightarrow \Delta}{\exists x A_x^a, \Gamma \rightarrow \Delta}$$

REMARK 1. To facilitate comparison with the systems of natural deduction, I have not taken $\sim$ as a primitive sign (which is otherwise the more natural procedure in the case of the calculi of sequents). If that is done and $\wedge$ is omitted from the languages, one adds the two rules

$$\rightarrow \sim) \quad \frac{A, \Gamma \rightarrow \Delta}{\Gamma \rightarrow \Delta, \sim A} \qquad \sim \rightarrow) \quad \frac{\Gamma \rightarrow \Delta, A}{\sim A, \Gamma \rightarrow \Delta}$$

~~which now hold as derived rules.~~ The $\wedge$-rule is then omitted and in the case of intuitionistic logic is replaced by a rule allowing addition of a formula in the succedent, so-called *thinning*:

$$\frac{\Gamma \to}{\Gamma \to A}$$

*The equivalences between the calculi obtained in this way and the calculi defined above are easily seen.*

REMARK 2. As defined above, the sequents that occur in a proof never contain two occurrences of the same formula on the same side of the arrow. One may allow such repetitions and state a *contraction rule* for deleting repetious occurrences. Clause (i) in the definition of proof can also be replaced by: (i′) every top-sequent has the form $A \to A$. In that case, one has to add a rule for thinning in both the antecedent and the succedent.

## § 2. Connections between the calculi of sequents and the systems of natural deduction

The calculi of sequents can be understood as meta-calculi for the deducibility relation in the corresponding systems of natural deduction. A sequent

$$\Gamma \to A$$

can be interpreted as the statement

$$\Gamma \vdash A;$$

and a sequent

$$\Gamma \to A_1 \, A_2 \dots A_n$$

where $n > 1$ (which can occur only in classical logic) can be interpreted as the statement

$$\Gamma \cup \{\sim A_1, \sim A_2, \dots, \sim A_{n-1}\} \vdash A_n.$$

It can easily be seen that the calculi are sound under this interpretation, i.e. it holds: if the sequent $\Gamma \to A$ is provable in the calculus for minimal logic, then $\Gamma \vdash_M A$; and similarly for intuitionistic and classical logic. (Obviousy this holds for top-sequents, and if it holds for the premisses of an instance of one of the inference rules above, then it holds also for the consequence.)

REMARK. A proof in a calculus of sequents can be looked upon as an instruction on how to construct a corresponding natural deduction. This is particularly evident in the case of intuitionistic or minimal logic. A top-sequent then corresponds to a natural deduction consisting of just the formula that occur both in the antecedent and the succedent. As we go downwards in the proof in the calculus of sequents, we successively enlarge in two directions the corresponding natural deductions obtained at the upper levels. When we come to applications of succedent rules, we enlarge the corresponding natural deductions at the bottom, applying the corresponding I-rules; when we come to applications of antecedent rules, we usually enlarge the corresponding natural deductions at the top, applying the corresponding E-rules. The natural deduction obtained in this way can be seen to be normal; the formulas that occur as minimum formulas are the same as those that occur in both the antecedent and the succedent of a top-sequent in the proof in the calculus of sequents.

The proof in the calculus of sequents can in this way be said to prescribe (to some extent) a certain order in which a corresponding natural deduction can be constructed. This order is often irrelevant and is only partially mirrored in the corresponding natural deduction that results from the construction. Different proofs in the calculus of sequents may therefore correspond (in the way indicated) to the same natural deduction.

## § 3. Gentzen's Hauptsatz

A proof in a calculus of sequents is, so to say, always normal; it can be seen that except for $\wedge$, every formula that occurs in such a proof occurs also in one of the formulas in the end-sequent of the proof. The essential content of Gentzen's Hauptsatz is now that these calculi also are complete, so that everything provable can be proved in this normal form. This we can now easily prove.[1]

THEOREM. *If* $\Gamma \vdash A$ *in* C, I, *or* M, *then there is a proof of the sequent* $\Gamma \to A$ *in the corresponding calculus of sequents.*

[1] A semantical completeness proof (in the model-theoretical sense) is given in Kanger [1]. This is perhaps the easiest and most elegant proof of the Hauptsatz but is of course (in contrast to the one given here) not constructive. Similar proofs are given by Schütte [2] for a somewhat different system (se note 1 on p. 88) and by Hintikka [1] and Beth [1] (the latter proofs are not explicity related to any calculi).

I prove the theorem for all the three systems simultaneously. However, in the classical case, I omit the v- and ∃-rules and prove instead the following lemma (from which the theorem for that case immediately follows).

LEMMA. *Let* $\varDelta = \{\sim B_1, \sim B_2, ..., \sim B_n\} \subset \varGamma$ $(n \geqslant 0)$, *and let* $\varGamma^* = \varGamma - \varDelta$. *It* holds: If $\varGamma \vdash_{C'} A$, *then the sequent* $\varGamma^* \to B_1 B_2 ... B_n A$ *is provable in the calculus for classical logic.*

The proof proceeds by induction over the lengths of the normal deductions in the systems of natural deduction. The base is trivial. The induction step is divided into three cases. We assume that $\Pi$ is a normal deduction of $A$ from $\varGamma$ that has only pure parameters.

*Case I. The end-formula of* $\Pi$ *is the consequence of an I-rule.* We then apply the corresponding succedent rule in the calculi of sequents. As an illustrative example, we take the case when $\Pi$ has the form $((\Sigma_1/C)/B \supset C)$. In that case, $(\Sigma_1/C)$ is a deduction of $C$ from $\varGamma \cup \{B\}$ and is of shorter length than that of $\Pi$. By the induction assumption, there is a proof of $B, \varGamma \to C$ and in the classical case also of $B, \varGamma^* \to B_1 B_2 ... B_n C$. By application of the rule $\to \supset$, we get the desired proof.

*Case II. The end-formula of* $\Pi$ *is the consequence of a* $\wedge$*-rule.* The intuitionistic case is immediate, using the $\wedge$-rule in the calculus of sequents. In the classical case, $\Pi$ is of the form $((\Sigma_1/\wedge)/A)$, where $(\Sigma_1/\wedge)$ is a proof of $\wedge$ from $\varGamma \cup \{\sim A\}$. By the induction assumption, there is a proof of $\varGamma^* \to B_1 B_2 ... B_n A \wedge$; and, by application of the $\wedge$-rule, we get the desired proof.

*Case III. The end-formula of* $\Pi$ *is the consequence of an E-rule.* Let $\tau$ be a thread in $\Pi$ that contains no minor premiss. Since the I-part of $\tau$ is empty (cf. Theorem 3 in Ch. III and Corollary 3 in Ch. IV) and there is no minor premiss in $\tau$, no assumption can be discharged in $\tau$. Hence, the first formula occurring in $\tau$, say $C$, belongs to $\varGamma$ and is the major premiss of an E-rule. I again take the case when $C$ has the form $(C_1 \supset C_2)$ as an example (the other cases being simpler). $\Pi$ has then the form

$$\frac{\dfrac{\Sigma_1}{C_1} \quad C_1 \supset C_2}{\dfrac{(C_2)}{\Pi_2}}$$

Since no assumption is discharged at any formula occurrence in $\tau$, $C_1$ cannot depend on other assumptions than those on which the end-formula of $\Pi$ depends. Hence $(\Sigma_1/C_1)$ is a deduction of $C_1$ from $\Gamma$; furthermore, $\Pi_2$ is a deduction of $A$ from $\Gamma \cup \{C_2\}$. By the induction assumption, there are proofs of $\Gamma \rightarrow C_1$ and of $C_2, \Gamma \rightarrow A$, and thus, by application of the rule $\supset \rightarrow$, we get a proof of $\Gamma \rightarrow A$. In the classical case, there are similar proofs when the succedent also contains $B_1$, $B_2, ...,$ and $B_n$.

REMARK. Gentzen [3], too, proved the completeness of the calculi of sequents by proving them equivalent to other logical systems. For this purpose, he added to the collection of inference rules the following *cut-rule* (which has a different character from that of the other rules):

$$\frac{\Gamma_1 \rightarrow \Delta_1, A \qquad A, \Gamma_2 \rightarrow \Delta_2}{\Gamma_1 \cup \Gamma_2 \rightarrow \Delta_1 \cup \Delta_2}$$

His Hauptsatz was then the assertion that this rule holds as a derived rule in his calculi, which he proved by showing that every application of the cut-rule can be eliminated from a given proof. This fact is of course now a corollary of what has been shown above, since it holds trivially that if $\Gamma_1 \vdash A$ and $\Gamma_2 \cup \{A\} \vdash B$, then also $\Gamma_1 \cup \Gamma_2 \vdash B$ (e.g., replace the assumptions of the form $A$ in the deduction of $B$ from $\Gamma_2 \cup \{A\}$ with the deduction of $A$ from $\Gamma_1$, arranging it so that the proper parameters in the first deduction do no not occur in the second).[1]

---

[1] We can easily extend this appendix to cover also ramified second order logic. Concerning the Hauptsatz for simple second order logic, cf. Chapter V, § 3.—An extension of Gentzen's Hauptsatz to S4 is given by Curry [2]. A calculus of sequents for S5 for which the Hauptsatz is stated (but without proof) is given by Kanger [1]. Other calculi of sequents for S5 for which the Hauptsatz is proved are given by Ohnishi–Matsumoto [1] and Hacking [1]. A Hauptsatz for pure relevant implication (corresponding to $R_p$) is stated without proof by Kripke [2]. It is not known whether it can be extended to calculi of sequents corresponding to $R_M$ and $R_C$ (and their extensions by modal rules), and the results of Chapter VI and VII cannot immediately be carried over to the calculi of sequents by the method of this appendix.

# APPENDIX B

## ON A SET THEORY BY FITCH

### § 1. A demonstrably consistent set theory

Fitch [2] developed a certain "demonstrably consistent" set theory. Using the results of the previous chapters, we can throw some light upon this system. In this section, I define a certain system of natural deduction for set theory, called F; its relation to Fitch's system is then discussed in § 2.

I consider a *language* that contains exactly one descriptive constant, the 2-place predicate constant $\epsilon$. I write $t\epsilon u$ or $(t\epsilon u)$ to denote the formula consisting of $\epsilon$ followed by the terms $t$ and $u$ (and similary for other letters). In addition to the non-descriptive symbols in Chapter I, the language is to contain also the operator $\lambda$ for set abstraction. The *individual terms* and the *formulas* in the language are defined by a simultaneous induction (cf. Chapter V). To the three clauses defining the notion of formula in Chapter I, § 1, we add the following ones:

4) An individual parameter is an individual term.

5) If $A$ is a formula, then $\lambda x A^*$ is an individual term, where $A^*$ is $A$ or is obtained from $A$ by replacing a parameter with $x$.

The distinction between bound and free variables is the same as in Chapter V, § 1.

We state an introduction and an elimination rule for $\lambda$, indicated as follows:

$$\lambda\text{I)} \quad \frac{A_t^x}{t\,\epsilon\,\lambda x A} \qquad\qquad \lambda\text{E)} \quad \frac{t\,\epsilon\,\lambda x A}{A_t^x}$$

F [F$_{c'}$] is to be the system that is obtained from I [C′] by the addition of the $\lambda$I- and $\lambda$E-rule. A *quasi-deduction* in one of these systems is defined in the same way as a deduction was defined for Gentzen-type systems in general (Chapter I, § 2). A *deduction* is then defined as a quasi-deduction that is normal.

The theorems about the form of normal deductions in Chapter III and IV hold now if we omit what is said about subformulas. (We have not defined that notion for the language in question.)

It can be seen that the systems are consistent in the sense that $\lambda$ is not provable in them (see the proof of Corollary 3 in Ch. III).

In other words, the set-theoretical paradoxes are ruled out by the requirement that the deductions shall be normal. For example, consider Russell's paradox, for which we have the following quasi-proof, where we let $t$ stand for the term $\lambda x \sim (x \in x)$:

$$
\cfrac{
  \cfrac{
    t \in t \quad \cfrac{\overset{(1)}{t \in t}}{\sim (t \in t)}
  }{
    \cfrac{\lambda}{\sim (t \in t)}^{(1)}
  }
}{
  t \in t
}
\qquad
\cfrac{
  \cfrac{
    t \in t \quad \cfrac{\overset{(1)}{t \in t}}{\sim (t \in t)}
  }{
    \cfrac{\lambda}{\sim (t \in t)}^{(1)}
  }
}{}
\;\Big/\; \lambda
$$

It follows from the above that this quasi-proof of $\lambda$ cannot be transformed into a normal proof. We have thus an example of a system for which the inversion principle holds (also the $\lambda$-rules obviously satisfy the principle) and where we hence can remove any given maximum formula, but where it is impossible to remove all maximum formulas from certain deductions. If we consider successive reductions of the quasi-proof above, we will see that we will oscillate infinitely between $\supset$-reductions and $\lambda$-reductions.

REMARK. It appears from the works of Fitch that one can develop a considerable portion of ordinary mathematics in F somewhat amplified. In spite of this fact and the demonstrable consistency of F, the system has serious disadvantages. Thus, although $\supset E$ is a rule of the system, one cannot in general infer that $B$ is provable given that $A$ and $A \supset B$ are provable, since there may be only a quasi-proof of $B$. This renders investigations of the system rather difficult as it is not sufficient to derive the axioms of an ordinary mathematical theory in the system in order to conclude that also its theorems are provable in the system.

## § 2. The relation of the system F to Fitch's system

The system F is constructed along the same general lines as the system(s) in Fitch [2], but they are not identical. Fitch considers two alternative restrictions that may be put on the quasi-deductions in order to avoid contradictions. What he calls the simple restriction seems to imply in our terminology that in one and the same thread, a premiss of an I-rule must not precede the premiss of an E-rule.[1] This is an unnecessarily strong restriction, since the requirement that a deduction is to be normal only implies that the same holds with respect to paths.[2] Fitch's alternative restriction (called the special restriction) is not quite comparable to the requirement that a deduction is to be normal. The restriction that is put on the deductions in F has the advantage that one knows that deducibility is not affected as long as $\epsilon$ and $\lambda$ are not both present.[3] The restriction is in this way less *ad hoc*.

Fitch's system also contains a different kind of negation, which among other things does not allow the inference of $\sim A$ given a deduction of a contradiction from $A$. But this has actually nothing to do with the consistency of his system. In fact, intuitionistic negation can be defined in Fitch's system: Every formula $A$ is deducible from $\forall x(x \epsilon x)$ (by $\forall$E followed by $\lambda$E); hence we can let $\lambda$ be a formula $\forall x(x \epsilon x)$ and define $\sim$ as in Chapter I. Fitch's system thus contains the same theory of negation as F does, and, in this sense, for certain formulas, we can prove both the formula and its negation in Fitch's system (e.g., for the formula $t \epsilon t$ considered in § 1). As seen in § 1, the $\lambda_c$-rule can also be added without destroying the consistency of the system (in the sense that $\lambda$ still remains unprovable).

Fitch's concept of negation may however have an interest in itself. It corresponds partially to what we can call *constructible falsity*,[4] which we can describe briefly as follows. Add an additional logical constant $\neg$

---

[1] Since the proofs in Fitch's system are linear rather than tree-formed, comparisons are not immediate on this point.

[2] Fitch's restriction probably rules out many ordinary logical laws even when $\epsilon$ and $\lambda$ are not present. (E.g., it is not seen how $P_1 \supset (P_2 \supset ((P_1 \& P_2 \supset Q) \supset Q))$ can be proved in his system.)

[3] Fitch's special restriction seems to rule out the possibility of proving $(P \equiv (P \supset Q)) \supset Q$.

[4] Cf. Nelson [1].

to the languages of Chapter I, and let $\neg A$ be a formula if $A$ is a formula. For $\neg$ we state the following I- and E-rules:[1]

$$\neg\,\&\,\text{I)}\quad \frac{\neg\,A}{\neg\,(A\,\&\,B)}\qquad \frac{\neg\,B}{\neg\,(A\,\&\,B)}\qquad \neg\,\&\,\text{E)}\quad \frac{\neg\,(A\,\&\,B)\quad \overset{(\neg A)}{C}\quad \overset{(\neg B)}{C}}{C}$$

$$\neg\,\supset\,\text{I)}\quad \frac{A\quad \neg\,B}{\neg\,(A\supset B)}\qquad \neg\,\supset\,\text{E)}\quad \frac{\neg(A\supset B)}{A}\qquad \frac{\neg(A\supset B)}{\neg\,B}$$

$$\neg\,\neg\,\text{I)}\quad \frac{A}{\neg\,\neg\,A}\qquad \neg\,\neg\,\text{E)}\quad \frac{\neg\,\neg\,A}{A}$$

$$\neg\,\forall\text{I)}\quad \frac{\neg\,A_i^z}{\neg\,\forall x A}\qquad \neg\,\forall\text{E)}\quad \frac{\neg\,\forall x A\quad \overset{(\neg A_a^x)}{B}}{B}$$

As in Chapter I, § 1, the letters within parentheses are to indicate how assumptions are discharged by applications of these rules. The restrictions on the $\neg\,\forall$E-rule are as those for the $\exists$E-rule. We may also add similar rules for the constructible falsity of disjunctions and existential formulas, but $\vee$ and $\exists$ can also be defined in terms of $\neg$, &, and $\forall$ in such a way that the rules in question become derivable.

Constructible falsity has clearly a constructive character. If we add the rules above to the system for intuitionistic logic, we can see that $\neg A$ is provable only if we can prove the constructible falsity of certain subformulas of $A$. Consequently, the formula $\neg\,(P\,\&\,\neg\,P)$ is not provable. To wit, the inversion principle holds also for these rules, and the theorems on normal deductions can be extended to the system mentioned; the assertions just made then follow as corollaries.

Intuitionistic and constructible falsity can be connected by the additional rule:[2]

$$\frac{A\quad \neg A}{\lambda}$$

$\sim A$ then becomes deducible from $\neg A$ but the converse is not true in general.

---

[1] Fitch does not use the rules for $\neg\,\supset$.

[2] Fitch has a rule to the same effect.

# APPENDIX C

## NOTES ON SOME OTHER VARIANTS OF NATURAL DEDUCTION

### § 1. The origin of natural deduction

The first person to express the idea of constructing a system of natural deduction seems to have been Łukasiewics in seminars in 1926. He called attention to the fact that in informal mathematical reasoning, one does not proceed according to the principles of the then common logical systems of Frege, Russell, and Hilbert among others, drawing inferences from axioms (or theorems) with the help of (proper) inference rules. Instead, one uses most frequently the method of drawing inferences from assumptions. Łukasiewics suggested that one should try to formalize this kind of reasoning, and the first results in this direction was obtained by *Jáskowski* in these seminars. He presented the results also at the First Polish Mathematical Congress in Lwow 1929.

Deductions in this system of Jáskowski's consist of formulas written in a number of boxes, some of which could appear within others. A new assumption is marked by the introduction of a new box. The assumption is written as the first formula in this box, and below one writes the formulas that are inferred from this assumption. If the box appears inside another box, formulas that stand in the larger box may be repeated in the smaller box. When an assumption is discharged and one obtains a consequence that is independent of the assumption, one writes this consequence outside and immediately below the box. The technique is exemplified by the deduction at the top of the next page.

In 1934, Jáskowski published a revised version of this system as well as some other systems under the title "On the rules of suppositions in formal logic" (Jáskowski [1]). He now develops systems for (1) classical sentential logic, (2) the sentential logic of Kolmogoroff, (3)

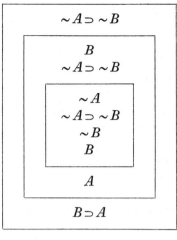

$$(\sim A \supset \sim B) \supset (B \supset A)$$

extended propositional logic (allowing quantifiers over propositional variables), and (4) a predicate logic in which the provable formulas are those classically valid in all domains including the empty one. I shall briefly describe systems (1), (2) and (4).

The languages of these systems are like those in Chapter I with some differences, the main ones of which are as follows: There are no individual parameters, and a formula is thus allowed to contain free variables. The logical constants are $\sim$, $\supset$, $\forall$. There are no descriptive symbols.

Instead of using the device with boxes explained above, Jáskowski now prefixes the formula occurrences in a deduction by strings of numerals; the numerals indicate the assumptions that the formula occurrences depend on. Let us call a string of numerals separated by commas—including the empty string—a *prefix*, and let $p$ and $q$ (sometimes with subscripts) refer to prefixes. Let us write $p \leqslant q$ to denote that $p$ agrees with an initial part of $q$ which is to be understood as including the case where $p = q$ and the case where $p$ is the empty prefix; let us say that $p$ *immediately precedes* $q$ if either $p$ is the empty prefix and $q$ is a numeral, or $p = q,\nu$ where $\nu$ is a numeral.

In a sequence $p_1A_1, p_2A_2, ..., p_nA_n$, we say that $p_iA_i$ is an *assumption* if $p_i$ is not empty and does not occur earlier in the sequence.

Somewhat changing Jáskowski's terminology, we can make the following definition:

$\mathcal{D}$ is a *deduction* in *Jáskowski's system for classical sentential logic* if and only if $\mathcal{D}$ is a sequence $p_1A_1, p_2A_2, ..., p_nA_n$ such that for each $i \leqslant n$ either

1) $p_iA_i$ is an assumption in $\mathcal{D}$; or

2) $A_i$ is obtained from $A_j$ and $A_k$ by $\supset$ E for some $j, k < i$ such that $p_j \lesssim p_i$ and $p_k \lesssim p_i$; or

3) $A_i = A_j \supset A_k$ for some $j, k < i$ such that $p_jA_j$ is an assumption in $\mathcal{D}$, $p_k \lesssim p_j$, and $p_i$ immediately precedes $p_j$; or

4) $\sim A_i = A_j$ for some $j < i$ such that $p_jA_j$ is an assumption in $\mathcal{D}$, $p_i$ immediately precedes $p_j$, and for some $k, m < i$ it holds that $A_k = \sim A_m$, $p_k \lesssim p_j$, and $p_m \lesssim p_j$.

$A$ is said to be deducible in this system from $\Gamma$ if there is a deduction $\mathcal{D}$ in the system ending with $pA$ for which it holds that if $q \lesssim p$ and $qB$ is an assumption in $\mathcal{D}$, then $B \in \Gamma$.

Below are two examples in this system of a proof of $(\sim A \supset \sim B) \supset (B \supset A)$ and of a deduction of $\sim \sim A \supset B$ from $A \supset \sim \sim B$:

| | | | | |
|---|---|---|---|---|
| 1 | $\sim A \supset \sim B$ | | 1 | $A \supset \sim \sim B$ |
| 1, 1 | $B$ | | 1, 1 | $\sim \sim A$ |
| 1, 1, 1 | $\sim A$ | | 1, 1, 1 | $\sim A$ |
| 1, 1, 1 | $\sim B$ | | 1, 1 | $A$ |
| 1, 1 | $A$ | | 1, 1 | $\sim \sim B$ |
| 1 | $B \supset A$ | | 1, 1, 2 | $\sim B$ |
| | $(\sim A \supset \sim B) \supset (B \supset A)$ | | 1, 1 | $B$ |
| | | | 1 | $\sim \sim A \supset B$ |

*Jaskowski's system for the sentential logic of Kolmogoroff* is obtained from the system above by changing "$\sim A_i = A_j$" in clause 4) to "$A_i = \sim A_j$".

As was said above, *Jaskowski's system for predicate logic* allows proofs only of formulas valid in all domains including the empty one. A *deduction* in this system is a sequence $\mathcal{D} = p_1\alpha_1, p_2\alpha_2, ..., p_n\alpha_n$ such that for each $i \leqslant n$, $\alpha_i$ is a formula or a variable and either

(I) $\alpha_i$ is a variable and $p_i$ is a non-empty prefix, neither of which occurs earlier in the sequence; or

(II) $\alpha_i$ is a formula for which one of the following clauses hold:

1) $p_i\alpha_i$ is an assumption in $\mathcal{D}$ (as defined before) and for every free variable $x$ in $\alpha_i$ there is a $j < i$ such that $x = \alpha_j$ and $p_j \lesssim p_i$;

2)–4) the same as for sentential logic;

5) there are $j$, $k < i$ such that $\alpha_j$ has the form $\forall x A$, $\alpha_k$ is a variable $y$, $x$ does not occur free in $A$ within the scope of a quantifier with $y$, $\alpha_i = A_y^x$, $p_j \lesssim p_i$, and $p_k \lesssim p_i$;

6) $\alpha_i = \forall \alpha_j \alpha_k$ for some $j$, $k < i$ such that $\alpha_j$ is a variable, $\alpha_k$ is a formula, $p_k \lesssim p_j$, and $p_i$ immediately precedes $p_j$.

$A$ is *provable* in this system if $A$ occurs without prefix as the last element of a deduction in the system.

Independently of this Polish development, *Gentzen* constructed in 1933 what he called "ein Kalkül des natürlichen Schliessens". It was published in 1934 in a paper with the title "Untersuchungen über das logische Schliessen" (Gentzen [3])[1], where he developed systems of natural deduction for classical and intuitionistic logic, essentially the ones described in Chapter I. (For some minor deviations see Remarks 1 and 2 in Chapter I, § 2)[2]

The minimal logic was introduced by *Johansson* [1] in 1937, and was stated by him in the form of a Gentzen-type system of natural deduction (as well as in the form of a calculus of sequents and in the form of a system of axiomatic type).[3]

## § 2. Variants of Gentzen-type systems

Various modifications of Gentzen's systems of natural deduction have been proposed. One variant, used in Gentzen [4], is to make

---

[1] A French translation of Gentzen's paper is given by Feys–Ladrière [1], supplemented by comments relating Gentzen's systems to the systems of Jáskowski and Johansson among others. Accounts of Gentzen's systems can also be found in Feys [1] and [2] and in Curry [1].

[2] Gentzen's description is more like the one in § 4 of Chapter I. A deduction is, according to him, to be supplemented by some marks that indicate the places at which the assumptions are discharged. (But these marks are not theoretically necessary. As the ∃E-rule is stated by Gentzen, however, it allows the discharge of assumptions of the same shape only, and one can then not, without such marks, always decide uniquely what assumptions the end-formula in a deduction depends on.)

[3] However, the part of minimal logic containing no logical constants besides ⊃ and ∼ coincides with the sentential logic of Kolmogoroff and was already developed in the form of a system of natural deduction by Jákowski as stated above.

explicit the formulas on which a formula occurrence in a deduction depends. This is done by replacing a formula occurrence $B$ that depends on the set $\Gamma$ by a sequent $A_1, A_2 \ldots A_n \rightarrow B$, where $A_1$, $A_2$, ..., $A_n$ is some ordering of the formulas in $\Gamma$ (cf. Appendix A). The deduction can then be understood as the transformation of sequents, which also in the classical case will have exactly one formula in the succedent. The assumptions become axioms of the forms $A \rightarrow A$. Applications of the I-rules become applications of the corresponding rules for introduction in the succedent in the calculi of sequents. Applications of the E-rules, however, do not become applications of rules for the introduction in the antecedent in the calculi of sequents. The deduction is thus still like a natural deduction rather than a proof in a calculus of sequents; in particular, the cut-rule holds trivially as a derived rule, and a theorem on normal deductions gets essentially the same form as in Chapter II and III.

This variant has the advantage that the deduction rules become proper inference rules, and that the definition of the deductions is simplified accordingly. It is then also possible to write the deduction in linear form instead of tree-form. When writing deductions in practice, however, it is cumbersome to repeat all the formulas in the antecedent at every step. Also on the meta-level (e.g. when defining reductions as in Chapter II, § 2), it is often more convenient to use Gentzen's original version.

To facilitate the writing of a deduction when using Gentzen's second version, *Feys–Ladrière* [1] suggest that every assumption be assigned a numeral and that the formulas in the antecedent be replaced by these numerals. (This device has been adopted in several text-books, e.g. in Suppes [1]).

To make deductions linear, the devices used by Jáskowski (or very similar devices) have often been employed. Jáskowski's original device is e.g. used by Fitch [2]. The part of a deduction that stands within a box is called a subordinate proof (by Fitch [2] and others) and is considered to be an item in the immediately larger box on a par with the formulas in the larger box.[1]

Linear deductions in the sense of Craig have not only the form of a sequence but satisfy the additional property that each formula in the

---

[1] The device of numerals used in Jáskowski's second formulation is replaced by Quine [2] with columns of stars (which is less clear if a new assumption is brought in after an earlier one has been discharged).

sequence implies the immediately succeeding one. A system for classical logic where the deductions are linear in this sense is given in *Craig* [1], and a theorem with partly the same aim as the theorem on normal deductions is proved for this system, using Gentzen's Hauptsatz for the classical calculus of sequents.

## § 3. Rules for existential instantiation

Because of their close correspondence to procedures common in informal reasoning, systems of natural deduction have often been used in textbooks for didactical purposes. Gentzen's and Jáskowski's systems are then usually modified in various ways. In particular, Gentzen's ∃E-rule is often replaced by a rule for so-called *existential instantiation* that is more symmetrical to the ∀I-rule.

One of the first systems of that kind is the one of *Quine*, published in 1950 in a paper "On natural deduction" and used in his book *Methods of logic* (Quine [1] and [2]). The simplification that may be obtained by replacing the ∃E-rule by a rule for existential instantiation is in Quine's system at the cost of other complications: His system contains rather cumbersome stipulations that require that certain variables are "flagged" and arranged in a certain order.

The stipulations concerning the ordering of the flagged variables are relaxed in *Gumin–Hermes* [1]. Although this considerably facilitates the construction of deductions, the stipulations are still rather complicated and artificial.

Another system that contains a rule for existential instantiation is developed by *Copi* [1] in his book *Symbolic logic*. Quine's stipulations are here replaced by some other restrictions on the two rules for introduction of ∀ and instantiation from ∃, which are much simpler in their wordings, but which are, on the other hand, very strong, sometimes forcing the deductions to be unnecessarily long. In Copi [2] the restrictions on the rule for introduction of ∀ are liberalized: $(A/\forall x A_x^y)$ is now an instance of that rule provided that (i) $A$ does not contain a free variable introduced by existential instantiation, and (ii) $y$ does not occur free in any assumption on which $A$ depends.[2]

---

[1] The same rules for existential instantiation and introduction of ∀ are stated by *Rosser* [1] as derived rules for a system of axiomatic type.

[2] As in most of the systems discussed here, there is no typographical distinction between bound and free variables. Obvious assumptions about $A_x^y$ are to be made.

See p. 6 for errata here.

Restriction (i), however, is still inconveniently strong. Consider e.g. the following deduction of $\exists z \forall y Pzy$ from $\exists z \forall x Pzx$ in Gentzen's system and the transformation of it into Copi's system, which violates his restriction (i):

<table>
<tr><td colspan="3" align="center">Gentzen's system:</td><td align="center">Copi's system:</td></tr>
<tr><td></td><td colspan="2" align="center">(1)</td><td></td></tr>
<tr><td></td><td colspan="2" align="center">$\forall x Pax$</td><td align="center">$\exists z \forall x Pzx$</td></tr>
<tr><td></td><td colspan="2" align="center">$Pab$</td><td align="center">$\forall x Pzx$</td></tr>
<tr><td>$\exists z \forall x Pzx$</td><td align="center">$\forall y Pay$</td><td rowspan="2">(1)</td><td align="center">$Pzx$</td></tr>
<tr><td colspan="2" align="center">$\forall y Pay$</td><td align="center">$\forall y Pzy$</td></tr>
<tr><td colspan="2" align="center">$\exists z \forall y Pzy$</td><td></td><td align="center">$\exists z \forall y Pzy$</td></tr>
</table>

A system of a somewhat different character is constructed by *Montague–Kalish* [1] and used in their text-book *Logic. Techniques of formal reasoning* (Kalish–Montague [1]). Applications of the improper inference rules $\supset$I, $\wedge_c$, and $\forall$I are in their system replaced by the following procedure. The formula that one wants to infer by such an application is first displayed and prefixed by "Show". The deduction that warrants the inference of this formula is then constructed immediately below. When completed, it is enclosed within a box, and the prefix "Show" is cancelled. Assumptions may be placed only immediately below a formula prefixed by "Show". Existential instantiation is allowed if the instatiated variable does not occur earlier in the deduction (in any form, not even bound in a formula occurrence prefixed by "Show"). The net effect of this is to allow about the same freedom of operation as in Gentzen's system.

Another device for existential elimination is introduced by *Borkowski–Słupecki* [1]. They use a new category of individual symbols that we may call ambiguous names. The terms are enriched by ambiguous names with subscripts consisting of possibly empty strings of individual parameters.[1] An instance of the rule for existential elimina-

---

[1] The same device is used by *Suppes* [1] in his book *Introduction to logic*, but the restrictions on his rules are more complicated. A simplification is given by *Lemmon* [1] resulting in the same rules as those of Borkowski–Słupecki. All these authors use free individual variables instead of parameters. This gives rise to certain difficulties pointed out by Lemmon [2], which however vanish when individual parameters are used as a in the formulation above.

tion has then the form $(\exists x A/A_\alpha^z)$ where $\alpha$ is a term consisting of an ambiguous name with a string of exactly all individual parameters in $A$ as subscript. Applications of this rule are to satisfy the restriction that the ambiguous name in question does not occur earlier in the deduction (the deductions are supposed to be linear). The other rules can be left as they are in Gentzen's system, or one can restrict the $\forall$I-rule so that it is required of an instance $(A/\forall x A_x^a)$ of this rule that $a$ does not occur in $A$ as subscript to an ambigous name. (The latter would have about the same effect as liberalizing Copi's restriction (i), replacing it by: (i') $A$ does not contain a free variable introduced by existential instantiation such that $y$ occurred in the formula obtained by that instantiation.)

Borkowski–Słupecki's device, too, gives the same freedom of operation as the rules in Gentzen's system. It seems to be essentially a matter of notation whether one chooses this device or Gentzen's device, where the conclusion that may be obtained by an application of the rule for existential instantiation is entered as an assumption which is later discharged at the appropriate time. However, Gentzen's system has the advantage that a formula occurrence in a deduction is always a logical consequence of the assumptions on which it depends. If one replaces the tree-form in Gentzen's system by a linear arrangement (e.g. one of those proposed by Jáskowski or Feys–Ladrière as described above), one seems indeed to have an eminently useful system also from a pedagogical viewpoint.

# BIBLIOGRAPHICAL REFERENCES

*JSL* = *The Journal of symbolic logic.*

ACKERMANN [1], WILHELM. Begründung einer strengen Implikation. *JSL*, vol. 21 (1956), pp. 113-128.

ANDERSON [1], ALAN ROSS. Comleteness theorems for the systems **E** of entailment and **EQ** of entailment with quantification. *Zeitschrift für mathematische Logik und Grundlagen der Mathematik*, vol. 6 (1960), pp. 201-206.

—— [2] Some open problems concerning the system **E** of entailment, *Acta Philosphica Phennica*, vol. 16, pp. 7-18. Helsinki 1963.

ANDERSON-BELNAP [1], A. R. and N. A modification of Ackermann's "rigorous implication" (abstract). *JSL*, vol. 23 (1958), pp. 457-458.

—— [2] The pure calculus of entailment. *JSL*, vol. 27 (1962), pp. 19-52.

BARCAN [1], RUTH. A functional calculus of first order based on strict implication. *JSL*, vol. 11 (1946), pp. 1-16.

BETH [1], E. W. *Semantic entailment and formal derivability.* Medelingen der Kon. Nederlandse Akad. van Wetensch., vol. 18, pp. 309-342. Amsterdam 1955.

BORKOWSKI-SŁUPECKI [1], L. and J. A logical system based on rules and its applications in teaching mathematical logic. *Studia Logica*, vol. 7 (1958), pp. 71-106.

CARNAP [1], RUDOLF. Modalities and quantification. *JSL*, vol. 11 (1946), pp. 33-64.

CHURCH [1], ALONZO. The weak theory of implication. *Kontrolliertes Denken (Festgabe zum 60. Geburtstag von Prof. W. Britzelmayr).* Munich 1951.

—— [2] *Introduction to mathematical logic.* Princeton 1956.

COPI [1], IRVING. *Symbolic logic.* New York 1954.

—— [2] Another variant of natural deduction. *JSL*, vol. 21 (1956), pp. 52-55.

CRAIG [1], WILLIAM. Linear reasoning. *JSL*, vol. 22 (1957), pp. 250-268.

CURRY [1], HASKELL. *A theory of formal deducibility.* (Notre Dame Mathematical lectures no. 6). Notre Dame 1950.

—— [2] The elimination theorem when modality is present. *JSL*, vol. 17 (1952), pp. 249-265.

—— [3] The inferential approach to logical calculus. *Logique et analyse*, vol. 3 (1960), pp. 119-136, and vol. 4 (1961), pp. 5-22.

FEYS [1], ROBERT. Les méthodes récentes de déduction naturelle. *Revue philosophique de Louvain*, vol. 44 (1946), pp. 370-400.

—— [2] Note complementaire sur les méthodes de déduction naturelle. *Ibid.*, vol. 45 (1947), pp. 60-72.

FEYS-LADRIÈRE [1], R. and J. Frech translation of Gentzen [3] supplemented with comments. French title: *Recherches sur la déduction loguique.* Paris 1955.

FITCH [1], FREDERIC. Intuitionistic modal logic with quantifiers. *Portugaliae mathematica*, vol. 7 (1948), pp. 113–118.

—— [2] *Symbolic logic*. New York 1952.

GENTZEN [1], GERHARD. Über die Existenz unabhängiger Axiomensystem zu unendlichen Satzsystemen. *Mathematische Annalen*, vol. 107 (1932), pp. 329–350.

—— [2] Über das Verhältnis zwischen intuitionistischen und klassischen Arithmetik. Manuscript set in type by Mathematische Annalen but not published ("Eingegangen am 15.3.1933").

—— [3] Untersuchungen über das logische Schliessen. *Mathematische Zeitschrift*, vol. 39 (1934), pp. 176–210.

—— [4] Die Widerspruchsfreiheit der reinen Zahlentheorie. *Mathematische Annalen*, vol. 112 (1936), pp. 493–565.

—— [5] Neue Fassung der Widerspruchsfreiheitsbeweise für die reine Zahlentheorie. *Forschungen zur Logik und zur Grundlegung der exakten Wissenschaften, Neue Folge*, Heft 4, pp. 19–44. Leipzig 1938.

GÖDEL [1], KURT. Zur intuitionistischen Arithmetik und Zahlentheorie. *Ergebnisse eines mathematischen Kolloquiums*, Heft 4 (for 1931–32, pub. 1933), pp. 34–38.

—— [2] Eine Interpretation des intuitionistischen Aussagenkalküls. *Ibid.*, pp. 39–40.

GUMIN-HERMES [1], H. and H. Die Soundness des Prädikatenkalküls auf der Basis der Quineschen Regeln. *Archiv für mathematische Logik und Grundlagenforschung*, vol. 2 (1954–1956), pp. 68–77.

HACKING [1], IAN. What is strict implication? *JSL*, vol. 28 (1963), pp. 51–71.

HARROP [1], R. On disjunctions and existential statements in intuitionistic systems of logic. *Mathematische Annalen*, vol. 132 (1956), pp. 347–361.

—— [2] Concerning formulas of the type $A \to B \vee C$, $A \to (Ex)B(x)$ in intuitionistic formal systems. *JSL*, vol. 25 (1960), pp. 27–32.

HENKIN [1], LEON. Banishing the rule of substitution for functional variables. *JSL*, vol. 18 (1953), pp. 201–208.

—— [2] An extension of the Craig-Lyndon interpolation theorem. *JSL*, vol. 28 (1963), pp. 201–216.

HERMES [1], HANS. Zum Inversionsprinzip der operativen Logik. *Constructivity in mathematics* (ed. A. Heyting), pp. 62–68. Amsterdam 1959.

HERTZ [1], PAUL. Über Axiomensysteme für beliebige Satzsysteme. *Mathematische Annalen*, vol. 87 (1922), pp. 246–269, and vol. 89 (1923), pp. 76–102.

HEYTING [1], AREND. *Intuitionism*. Amsterdam 1956.

HILBERT-BERNAYS [1], D. and P. *Grundlagen der Mathematik, I*. Berlin 1934.

HINTIKKA [1], JAAKKO. Form and content in quantification theory. *Acta Philosphica Fennica*, vol. 8, pp. 11–55. Helsinki 1955.

HORN [1], ALFRED. The separation theorem of intuitionistic propositional calculus. *JSL*, vol. 27 (1962), pp. 391–399.

JAŚKOWSKI [1], STANISŁAW. On the rules of suppositions in formal logic. *Studia logica*, no. 1. Warsaw 1934.

JOHANSSON [1], INGEBRIGT. Der Minimalkalkül, ein reduzierter intuitionistischer Formalismus. *Compositio mathematica*, vol. 4 (1936) ,pp. 119–136.

KALISH-MONTAGUE [1], D. and R. *Logic, techniques of formal reasoning.* New York 1964.

KANGER [1], STIG. *Provability in logic.* Stockholm 1957.

KLEENE [1], STEPHEN. *Introduction to matemathematics.* Amsterdam 1952.

—— [2] Disjunction and existence under implication in elementary intuitionistic formalisms. *JSL*, vol. 27 (1962), pp. 11–18.

—— [3] An addendum. *Ibid.*, vol. 28 (1963), pp. 154–156.

KNEALE [1], WILLHELM. The province of logic. *Contemporary Brittish philosophy* (ed. H. D. Lewis), third series, pp. 235–261. Aberdeen 1956.

KRIPKE [1], SAUL. A completeness theorem in modal logic. *JSL*, vol. 24 (1959), pp. 1–15.

—— [2] The problem with entailment (abstract). *JSL*, vol. 24 (1959), p. 324.

LEMMON [1]. Quantifier rules and natural deduction. Mind, vol. 70 (1961), pp. 235–238.

—— [2] Existential specification in natural deduction (abstract). *JSL*, vol. 28 (1963), p. 263.

LEWIS-LANGFORD [1], C. I. and C. H. *Symbolic logic.* New York 1932.

LÖB [1], M. H. Cut elimination in type-theory (abstract). *JSL*, vol. 29 (1964), p. 220.

LORENZEN, [1], PAUL. Konstruktive Begründung der Mathematik. *Mathematische Zeitschrift*, vol. 53 (1950), pp. 162–201.

—— [2] *Einführung in die operative Logik und Mathematik.* Berlin 1955.

LYNDON [1], ROGER. An interpolation theorem in the predicate calculus. *Pacific journal of mathematics*, vol. 9 (1959), pp. 129–142.

McKINSEY [1], J. C. C. Proof of the independence of the primitive symbols of Heyting's calculus of propositions. *JSL*, vol. 4 (1939), pp. 155–158.

MCKINSEY-TARSKI [1], J. and A. Some theorems about the sentential calculi of Lewis and Heyting. *JSL*, vol. 13 (1948), pp. 1–15.

MONTAGUE-KALISH [1], R. and D. Remarks on descriptions and natural deduction. *Archiv für mathematische Logik und Grundlagenforschung*, vol. 3 (1957), pp. 50–64.

NELSON [1], DAVID. Constructible falsity. *JSL*, vol. 14 (1949), pp. 16–26.

OHNISHI-MATSUMOTO [1], M. and K. Gentzen method in modal calculi, II. *Osaka mathematical journal*, vol. 11 (1959), pp. 115–120.

PRIOR [1], ARTHUR. Modality and quantification in S5. *JSL*, vol. 21 (1956), pp. 60–62.

QUINE [1], W. V. On natural deduction. *JSL*, vol. 15 (1950), pp. 93–102.

—— [2] *Methods of logic.* New York 1950.

ROSSER [1], BERKLEY. *Logic for mathematicians.* New York 1953.

RUSSELL [1], BERTRAND. *The principles of mathematics.* Cambridge 1903.

SCHÜTTE [1], KURT. Schlussweisen-Kalkül der Prädikatenlogik. *Mathematische Annalen*, vol. 122 (1950–51), pp. 47–65.

—— [2] Ein System des verknüpfenden Schliessens. *Archiv für mathematische Logik und Grundlagenforschung*, vol. 2 (1954–56), pp. 55–67.

—— [3] Der Interpolationssatz der intuitionistischen Prädikatenlogik. *Mathematische Annalen*, vol. 148 (1962), pp. 192–200.

SUPPES [1], PATRIK. *Introduction to logic*. Princeton 1957.

TAKEUTI [1], GAISI. On a generalized logic calculus. *Japanese journal of mathematics*, vol. 23 (1953), pp. 39–96. Errata, *Ibid.*, vol. 24 (1954), 149–156.

WAJSBERG [1], MORDECHAJ. Untersuchungen über den Aussagenkalkül von A. Heyting. *Viadomości Matematyczne*, vol. 46 (1938), pp. 45–101.

# INDEX

# INDEX OF SYMBOLS

# A CATALOG OF SELECTED
# DOVER BOOKS
## IN SCIENCE AND MATHEMATICS

# Mathematics

FUNCTIONAL ANALYSIS (Second Corrected Edition), George Bachman and Lawrence Narici. Excellent treatment of subject geared toward students with background in linear algebra, advanced calculus, physics and engineering. Text covers introduction to inner-product spaces, normed, metric spaces, and topological spaces; complete orthonormal sets, the Hahn-Banach Theorem and its consequences, and many other related subjects. 1966 ed. 544pp. 6⅛ x 9¼. 0-486-40251-7

ASYMPTOTIC EXPANSIONS OF INTEGRALS, Norman Bleistein & Richard A. Handelsman. Best introduction to important field with applications in a variety of scientific disciplines. New preface. Problems. Diagrams. Tables. Bibliography. Index. 448pp. 5⅜ x 8½. 0-486-65082-0

VECTOR AND TENSOR ANALYSIS WITH APPLICATIONS, A. I. Borisenko and I. E. Tarapov. Concise introduction. Worked-out problems, solutions, exercises. 257pp. 5⅝ x 8¼. 0-486-63833-2

AN INTRODUCTION TO ORDINARY DIFFERENTIAL EQUATIONS, Earl A. Coddington. A thorough and systematic first course in elementary differential equations for undergraduates in mathematics and science, with many exercises and problems (with answers). Index. 304pp. 5⅜ x 8½. 0-486-65942-9

FOURIER SERIES AND ORTHOGONAL FUNCTIONS, Harry F. Davis. An incisive text combining theory and practical example to introduce Fourier series, orthogonal functions and applications of the Fourier method to boundary-value problems. 570 exercises. Answers and notes. 416pp. 5⅜ x 8½. 0-486-65973-9

COMPUTABILITY AND UNSOLVABILITY, Martin Davis. Classic graduate-level introduction to theory of computability, usually referred to as theory of recurrent functions. New preface and appendix. 288pp. 5⅜ x 8½. 0-486-61471-9

ASYMPTOTIC METHODS IN ANALYSIS, N. G. de Bruijn. An inexpensive, comprehensive guide to asymptotic methods–the pioneering work that teaches by explaining worked examples in detail. Index. 224pp. 5⅜ x 8½ 0-486-64221-6

APPLIED COMPLEX VARIABLES, John W. Dettman. Step-by-step coverage of fundamentals of analytic function theory–plus lucid exposition of five important applications: Potential Theory; Ordinary Differential Equations; Fourier Transforms; Laplace Transforms; Asymptotic Expansions. 66 figures. Exercises at chapter ends. 512pp. 5⅜ x 8½. 0-486-64670-X

INTRODUCTION TO LINEAR ALGEBRA AND DIFFERENTIAL EQUATIONS, John W. Dettman. Excellent text covers complex numbers, determinants, orthonormal bases, Laplace transforms, much more. Exercises with solutions. Undergraduate level. 416pp. 5⅜ x 8½. 0-486-65191-6

RIEMANN'S ZETA FUNCTION, H. M. Edwards. Superb, high-level study of landmark 1859 publication entitled "On the Number of Primes Less Than a Given Magnitude" traces developments in mathematical theory that it inspired. xiv+315pp. 5⅜ x 8½. 0-486-41740-9

CALCULUS OF VARIATIONS WITH APPLICATIONS, George M. Ewing. Applications-oriented introduction to variational theory develops insight and promotes understanding of specialized books, research papers. Suitable for advanced undergraduate/graduate students as primary, supplementary text. 352pp. 5⅜ x 8½.
0-486-64856-7

COMPLEX VARIABLES, Francis J. Flanigan. Unusual approach, delaying complex algebra till harmonic functions have been analyzed from real variable viewpoint. Includes problems with answers. 364pp. 5⅜ x 8½.
0-486-61388-7

AN INTRODUCTION TO THE CALCULUS OF VARIATIONS, Charles Fox. Graduate-level text covers variations of an integral, isoperimetrical problems, least action, special relativity, approximations, more. References. 279pp. 5⅜ x 8½.
0-486-65499-0

COUNTEREXAMPLES IN ANALYSIS, Bernard R. Gelbaum and John M. H. Olmsted. These counterexamples deal mostly with the part of analysis known as "real variables." The first half covers the real number system, and the second half encompasses higher dimensions. 1962 edition. xxiv+198pp. 5⅜ x 8½. 0-486-42875-3

CATASTROPHE THEORY FOR SCIENTISTS AND ENGINEERS, Robert Gilmore. Advanced-level treatment describes mathematics of theory grounded in the work of Poincaré, R. Thom, other mathematicians. Also important applications to problems in mathematics, physics, chemistry and engineering. 1981 edition. References. 28 tables. 397 black-and-white illustrations. xvii + 666pp. 6⅛ x 9¼.
0-486-67539-4

INTRODUCTION TO DIFFERENCE EQUATIONS, Samuel Goldberg. Exceptionally clear exposition of important discipline with applications to sociology, psychology, economics. Many illustrative examples; over 250 problems. 260pp. 5⅜ x 8½.
0-486-65084-7

NUMERICAL METHODS FOR SCIENTISTS AND ENGINEERS, Richard Hamming. Classic text stresses frequency approach in coverage of algorithms, polynomial approximation, Fourier approximation, exponential approximation, other topics. Revised and enlarged 2nd edition. 721pp. 5⅜ x 8½. 0-486-65241-6

INTRODUCTION TO NUMERICAL ANALYSIS (2nd Edition), F. B. Hildebrand. Classic, fundamental treatment covers computation, approximation, interpolation, numerical differentiation and integration, other topics. 150 new problems. 669pp. 5⅜ x 8½. 0-486-65363-3

THREE PEARLS OF NUMBER THEORY, A. Y. Khinchin. Three compelling puzzles require proof of a basic law governing the world of numbers. Challenges concern van der Waerden's theorem, the Landau-Schnirelmann hypothesis and Mann's theorem, and a solution to Waring's problem. Solutions included. 64pp. 5⅜ x 8½.
0-486-40026-3

THE PHILOSOPHY OF MATHEMATICS: AN INTRODUCTORY ESSAY, Stephan Körner. Surveys the views of Plato, Aristotle, Leibniz & Kant concerning propositions and theories of applied and pure mathematics. Introduction. Two appendices. Index. 198pp. 5⅜ x 8½. 0-486-25048-2

INTRODUCTORY REAL ANALYSIS, A.N. Kolmogorov, S. V. Fomin. Translated by Richard A. Silverman. Self-contained, evenly paced introduction to real and functional analysis. Some 350 problems. 403pp. 5⅜ x 8½. 0-486-61226-0

APPLIED ANALYSIS, Cornelius Lanczos. Classic work on analysis and design of finite processes for approximating solution of analytical problems. Algebraic equations, matrices, harmonic analysis, quadrature methods, much more. 559pp. 5⅜ x 8½. 0-486-65656-X

AN INTRODUCTION TO ALGEBRAIC STRUCTURES, Joseph Landin. Superb self-contained text covers "abstract algebra": sets and numbers, theory of groups, theory of rings, much more. Numerous well-chosen examples, exercises. 247pp. 5⅜ x 8½. 0-486-65940-2

QUALITATIVE THEORY OF DIFFERENTIAL EQUATIONS, V. V. Nemytskii and V.V. Stepanov. Classic graduate-level text by two prominent Soviet mathematicians covers classical differential equations as well as topological dynamics and ergodic theory. Bibliographies. 523pp. 5⅜ x 8½. 0-486-65954-2

THEORY OF MATRICES, Sam Perlis. Outstanding text covering rank, nonsingularity and inverses in connection with the development of canonical matrices under the relation of equivalence, and without the intervention of determinants. Includes exercises. 237pp. 5⅜ x 8½. 0-486-66810-X

INTRODUCTION TO ANALYSIS, Maxwell Rosenlicht. Unusually clear, accessible coverage of set theory, real number system, metric spaces, continuous functions, Riemann integration, multiple integrals, more. Wide range of problems. Undergraduate level. Bibliography. 254pp. 5⅜ x 8½. 0-486-65038-3

MODERN NONLINEAR EQUATIONS, Thomas L. Saaty. Emphasizes practical solution of problems; covers seven types of equations. ". . . a welcome contribution to the existing literature...."–*Math Reviews*. 490pp. 5⅜ x 8½. 0-486-64232-1

MATRICES AND LINEAR ALGEBRA, Hans Schneider and George Phillip Barker. Basic textbook covers theory of matrices and its applications to systems of linear equations and related topics such as determinants, eigenvalues and differential equations. Numerous exercises. 432pp. 5⅜ x 8½. 0-486-66014-1

LINEAR ALGEBRA, Georgi E. Shilov. Determinants, linear spaces, matrix algebras, similar topics. For advanced undergraduates, graduates. Silverman translation. 387pp. 5⅜ x 8½. 0-486-63518-X

ELEMENTS OF REAL ANALYSIS, David A. Sprecher. Classic text covers fundamental concepts, real number system, point sets, functions of a real variable, Fourier series, much more. Over 500 exercises. 352pp. 5⅜ x 8½. 0-486-65385-4

SET THEORY AND LOGIC, Robert R. Stoll. Lucid introduction to unified theory of mathematical concepts. Set theory and logic seen as tools for conceptual understanding of real number system. 496pp. 5⅜ x 8¼. 0-486-63829-4

TENSOR CALCULUS, J.L. Synge and A. Schild. Widely used introductory text covers spaces and tensors, basic operations in Riemannian space, non-Riemannian spaces, etc. 324pp. 5⅜ x 8¼. 0-486-63612-7

ORDINARY DIFFERENTIAL EQUATIONS, Morris Tenenbaum and Harry Pollard. Exhaustive survey of ordinary differential equations for undergraduates in mathematics, engineering, science. Thorough analysis of theorems. Diagrams. Bibliography. Index. 818pp. 5⅜ x 8½. 0-486-64940-7

INTEGRAL EQUATIONS, F. G. Tricomi. Authoritative, well-written treatment of extremely useful mathematical tool with wide applications. Volterra Equations, Fredholm Equations, much more. Advanced undergraduate to graduate level. Exercises. Bibliography. 238pp. 5⅜ x 8½. 0-486-64828-1

FOURIER SERIES, Georgi P. Tolstov. Translated by Richard A. Silverman. A valuable addition to the literature on the subject, moving clearly from subject to subject and theorem to theorem. 107 problems, answers. 336pp. 5⅜ x 8½. 0-486-63317-9

INTRODUCTION TO MATHEMATICAL THINKING, Friedrich Waismann. Examinations of arithmetic, geometry, and theory of integers; rational and natural numbers; complete induction; limit and point of accumulation; remarkable curves; complex and hypercomplex numbers, more. 1959 ed. 27 figures. xii+260pp. 5⅜ x 8½. 0-486-63317-9

POPULAR LECTURES ON MATHEMATICAL LOGIC, Hao Wang. Noted logician's lucid treatment of historical developments, set theory, model theory, recursion theory and constructivism, proof theory, more. 3 appendixes. Bibliography. 1981 edition. ix + 283pp. 5⅜ x 8½. 0-486-67632-3

CALCULUS OF VARIATIONS, Robert Weinstock. Basic introduction covering isoperimetric problems, theory of elasticity, quantum mechanics, electrostatics, etc. Exercises throughout. 326pp. 5⅜ x 8½. 0-486-63069-2

THE CONTINUUM: A CRITICAL EXAMINATION OF THE FOUNDATION OF ANALYSIS, Hermann Weyl. Classic of 20th-century foundational research deals with the conceptual problem posed by the continuum. 156pp. 5⅜ x 8½. 0-486-67982-9

CHALLENGING MATHEMATICAL PROBLEMS WITH ELEMENTARY SOLUTIONS, A. M. Yaglom and I. M. Yaglom. Over 170 challenging problems on probability theory, combinatorial analysis, points and lines, topology, convex polygons, many other topics. Solutions. Total of 445pp. 5⅜ x 8½. Two-vol. set. Vol. I: 0-486-65536-9 Vol. II: 0-486-65537-7

Paperbound unless otherwise indicated. Available at your book dealer, online at **www.doverpublications.com**, or by writing to Dept. GI, Dover Publications, Inc., 31 East 2nd Street, Mineola, NY 11501. For current price information or for free catalogues (please indicate field of interest), write to Dover Publications or log on to **www.doverpublications.com** and see every Dover book in print. Dover publishes more than 500 books each year on science, elementary and advanced mathematics, biology, music, art, literary history, social sciences, and other areas.